TECHNICAL
REPORT

T0302554

An Economic Development Architecture for New Orleans

Kevin F. McCarthy

Sponsored by the Horizon Initiative

RAND GULF STATES POLICY INSTITUTE

This research was sponsored by the Horizon Initiative and was conducted within the RAND Gulf States Policy Institute (RGSPI)—a partnership involving the RAND Corporation and seven universities in the Gulf of Mexico coastal region.

Library of Congress Cataloging-in-Publication Data

McCarthy, Kevin F., 1945-
An economic development architecture for New Orleans / Kevin F. McCarthy.
p. cm.
Includes bibliographical references.
ISBN 978-0-8330-4324-5 (pbk. : alk. paper)
1. Economic development—Louisiana—New Orleans. 2. Louisiana—Economic policy. 3. Hurricane Katrina, 2005. I. Title.

HD77.5.L8M33 2008
338.9763'35—dc22

2007052737

Published 2008 by the RAND Corporation
1776 Main Street, P.O. Box 2138, Santa Monica, CA 90407-2138
1200 South Hayes Street, Arlington, VA 22202-5050
4570 Fifth Avenue, Suite 600, Pittsburgh, PA 15213-2665
RAND URL: http://www.rand.org/
To order RAND documents or to obtain additional information, contact
Distribution Services: Telephone: (310) 451-7002;
Fax: (310) 451-6915; Email: order@rand.org

Preface

In response to the devastating effects of Hurricane Katrina and the subsequent levee failures on New Orleans and its economy, the Horizon Initiative, a private-sector organization, was formed in 2006 to expedite the city's recovery with a particular focus on economic recovery. Horizon, in collaboration with the New Orleans City Council, has been evaluating alternative organizational structures and strategies to revitalize the city's economy. To assist in this effort, Horizon contracted with the RAND Gulf States Policy Institute (RGSPI) to provide recommendations as to the most effective organizational and strategic approaches to accomplish Horizon's objective. This report was produced in response to that challenge and aims to provide an organizational architecture for New Orleans's economic revitalization. As such, it specifically addresses three questions:

- What strategies should the effort employ?
- How should the effort be organized?
- How should priorities be assigned, and how should the approach be implemented?

This report is not a detailed redevelopment plan. The content and approach of a specific economic plan or strategy are essential for putting it into action, and such issues are beyond the scope of this report. What this research does is provide guidelines for evaluating both alternative organizational structures and the many economic development proposals being put forward for New Orleans's economic recovery.

In discussing the revitalization of New Orleans, this report draws a distinction between economic development and urban redevelopment programs. Economic development programs are designed to stimulate the economy by increasing the growth of employment and income of the city and the region as a whole. Urban redevelopment programs, on the other hand, are designed to improve the physical, social, and economic status of neighborhoods within the city. Although both sets of efforts are important (and at times related, since urban redevelopment is often an important component of increasing an area's attractiveness to new employers and residents), they have distinct goals and methods. Moreover, given the decline in New Orleans's economy prior to Katrina, simply restoring the region's economy to its prestorm condition will not serve the region's long-term economic needs. Indeed, without a healthy economy, the region's neighborhoods will all suffer. Much of the planning effort that has been undertaken in New Orleans since Katrina has focused more on urban redevelopment rather than economic development.

The RAND Gulf States Policy Institute

RGSPI is a collaboration among RAND and seven universities (Jackson State University, Tulane University, Tuskegee University, University of New Orleans, University of South Alabama, University of Southern Mississippi, and Xavier University) to develop a long-term vision and strategy to help build a better future for Louisiana, Mississippi, and Alabama in the wake of hurricanes Katrina and Rita. RGSPI's mission is to support a safer, more equitable, and more prosperous future for the Gulf of Mexico region by providing officials from the government, nonprofit organizations, and the private sector with relevant policy analysis of the highest caliber.

RGSPI is housed at the RAND Corporation, an international nonprofit research organization with a reputation for rigorous and objective analysis and effective solutions.

For additional information about RGPSI, contact its director:

George Penick
RAND Gulf States Policy Institute
P.O. Box 3788
Jackson, MS 39207
601-797-2499
George_Penick@rand.org

For a profile of RGSPI, see http://www.rand.org/about/katrina.html. More information about RAND is available at http://www.rand.org.

Contents

Figure and Tables

Figure

Tables

Summary

In August 2005, Hurricane Katrina, followed by multiple levee failures, devastated New Orleans and other parts of the U.S. Gulf Coast, inflicting major damage to commercial property, infrastructure, and housing. The failure of the levees and the flooding of New Orleans caused enormous damage and disruption to the city, its people, and its economy. Recovering from a disaster of this magnitude poses a major challenge to the city, the state, and the nation. The complexity of this challenge is compounded by the fact that New Orleans's population and economy had been lagging for several decades before Katrina. Although recovery is under way, it has been proceeding more slowly than the city's residents, local public and private leaders, and other stakeholders had hoped.

In response to this situation, the Horizon Initiative, a private-sector organization, formed to expedite the city's recovery, asked the RAND Gulf States Policy Institute (RGSPI) to provide recommendations as to the most effective organizational and strategic approaches to revitalizing the city's economy. Specifically, RGSPI was asked, first, to identify the best practices that other cities have used to foster economic development (in terms of both how they organize their efforts and the approaches they use); second, to describe how these practices might be applied to New Orleans; and third, to recommend actions that might be needed to facilitate the application of those practices in New Orleans. Consistent with the request by the Horizon Initiative, the overall architecture reflected in our recommendations addresses both how the organizational structure should be configured and how it should strategically focus its efforts.

The study that produced this report was conducted in three phases. The first was an examination of the development experiences of 17 other cities; the second was a review of studies of New Orleans's pre- and post-Katrina economic infrastructure and its prior development efforts; and the third involved interviews with a range of individuals from the city's public and private sectors.

Lessons from Other Cities

Our review of the economic development efforts of other cities revealed several important lessons:

- A successful economic development program requires a comprehensive design, an appropriate organization, and an effective implementation plan.
- A clear and comprehensive design starts with a vision of what the development authority and community are trying to accomplish, and includes both an appropriate plan or

approach to achieve that vision, including the identification of a set of industries to target with the strategy, and a commitment to make the required improvements to the area's economic infrastructure that will promote the growth of those target industries.
- Organizing for success is as important as the strategy used to achieve it. An effective organization requires coordination of the activities of the parties involved in the effort, with a clear division of roles and responsibilities among those parties.
- An effective division of labor is best served by establishing separate bodies to perform three key functions: a governing board to set policy, an executive staff to carry out that policy and coordinate activities, and a resource group to provide expertise as needed.
- The implementation of the strategy requires setting priorities and goals, as well as identifying metrics to evaluate progress in meeting those goals.

Overall Architecture

This review of the experience of a range of different metropolitan regions leads to our first recommendation for the overall architecture of New Orleans's development effort.

Recommendation One: New Orleans should build its economic development program in three phases: a design phase, an organizational phase, and an implementation phase.

Given critiques of New Orleans's prior development efforts, the controversy that surrounding the various neighborhood redevelopment plans, and the city's reputation for cronyism, we make a second recommendation for the overall architecture of New Orleans's development effort.

Recommendation Two: The process of developing and implementing the city's economic development program should be both transparent (open to public scrutiny) and inclusive (incorporating input from a diverse range of stakeholders).

Design Phase

The design phase of an economic development program typically includes four basic elements: first, a vision statement articulating the desired characteristics of the future economy; second, an assessment of the city's economic infrastructure that highlights its strengths and weaknesses; third, a description of the strategic focus of the effort, including its principal goals and targets; and, finally, a list of specific improvements to that infrastructure that are viewed as central to the plan.

Although vision statements are often cast in very general terms, the process of drafting them is important as a mechanism to publicize the development effort, involve a wide range of community stakeholders, and develop support for the economic development program. This leads to our third recommendation.

Recommendation Three: The vision statement for a New Orleans economic development effort should involve consultation with a wide range of stakeholders to reflect the diversity of their views and to make them aware of and vested in the development effort.

Our review of New Orleans's economic base and its current assets revealed that the city's industrial base has been concentrated around four industrial clusters: a maritime cluster, consisting of shipping (the port) and shipbuilding and repair; an oil and gas cluster, including production services and the petrochemical industry; a tourism cluster related to the city's unique facilities, festivals, and attractions; and a biomedical cluster that consists of health services and biomedical research industries. These clusters reflect the city's unique comparative advantages, most of which (e.g., its location on the Mississippi River, the nation's major north-south water artery; its extensive port facilities; its proximity to the nation's major oil and gas reserves in the Gulf of Mexico; its unique tourist attractions) will remain major advantages for New Orleans despite Katrina. On the other hand, the city's biomedical cluster, while based on a particularly strong set of facilities and a dominant position in the U.S. Gulf Coast region before Katrina, suffered major damage in the aftermath of the hurricane, as well as a substantial loss of its critical professional staff, and it will need to be repaired if it is to return to pre-Katrina capacity.

In addition, New Orleans has notable strengths in food processing, aerospace (the NASA Michaud facility), arts and entertainment (music and film production), and higher education, as well as promising clusters in such high-technology areas as information, biotech, and alternative energy. Finally, although devastated by Katrina, the small-business sector was a critical element of the city's economic infrastructure that provided 40 percent of the city's total employment before the storm and significant opportunities for minority entrepreneurs.

The review of city's economic base leads to our next recommendation.

Recommendation Four: The city's economic development efforts should be targeted at three groups of industrial clusters. The first group consists of three core industries that have made or are making significant recovery post-Katrina, namely the maritime, oil and gas, and tourism industries. The second group consists of three groups of industries or sectors that are vital to the city's economy but have been slower to recover from the damage caused by the storm—namely, the biomedical and food-processing industries and the city's small-business sector. These will require case-by-case evaluation. The third group consists of three clusters with solid pre-Katrina bases in the city and promising growth potential—namely, the arts and entertainment, information technology, and energy technology clusters.

New Orleans's economic infrastructure has several weaknesses, many of which predated Katrina but have been exacerbated by the damage done by the storm and its aftermath. Probably the single most important physical infrastructure problem is the uncertain state of the levees and the flood-control system protecting the city from future storms. Despite the repairs that have been made, the uncertainty surrounding the city's protection from future storms discourages much-needed investment, both by the city's existing businesses and residents and by potential new businesses and residents. In addition, the city's physical infrastructure needs major upgrading and repair, not just to such basic facilities as streets, transportation networks, and utilities, but especially to the port and the city's biomedical facilities.

Other major problems for the city's economic infrastructure include the quantity and quality of its labor force and the city's relationship with the business sector. Members of the city's business community whom we interviewed repeatedly complained about the difficulties they faced in dealing with city government, primarily the complexity of and delays in the permitting and licensing processes and the lack of a service orientation among city employees. This situation will pose special problems for the small businesses that will need special assistance to get reestablished.

Finally, the city must upgrade the quality of life for its residents and refurbish New Orleans's image. Despite its many charms, post-Katrina New Orleans is not an easy place to live. To remedy this situation, the city will need to address a rising crime problem, upgrade the quality of its schools and health facilities, provide a sufficient supply of safe and affordable housing, and restore basic services to its rebuilding neighborhoods. The New Orleans brand (the city's national image), which suffered pre-Katrina from perceptions of corruption and cronyism, has been further tarnished by the slow pace of recovery and additional indictments of city officials. Restoring the city's image is essential to securing the additional federal aid that the city needs and encouraging investment by business.

These weaknesses lead to our next recommendation.

Recommendation Five: New Orleans needs to recognize the importance of beginning its economic development programming by addressing the weaknesses in its economic infrastructure. Priority should be given to repairing and upgrading the city's infrastructure (e.g., water and sewers, public transportation), expanding the supply and improving the quality of the city's labor force, improving the quality of the interaction between the city and the business community, upgrading the quality of life for the city's residents, and refurbishing the city's national image.

Other cities typically choose among three different general strategic approaches to economic development—each with a distinct target, selling points, and marketing technique. The first approach aims to recruit new industries by emphasizing low costs, tax incentives and rebates, and a business-friendly environment. It aggressively markets these advantages to firms located in other areas. The second approach focuses on retaining and supporting the growth of industries and firms already located in the area by highlighting an improved physical infrastructure, an abundant supply of qualified labor, and government policies that are supportive of local industries. It then focuses on making the needed internal improvements to the city's economic infrastructure. The third approach targets new potential growth industries, with a particular focus on high-tech industries. Its major selling points are strong education and research sectors and a quality of life that attracts highly skilled, relatively "footloose" workers. It then focuses on building up its educational and research capabilities and a quality of life that will attract a highly skilled labor force on the assumption that, if the area can attract such labor, the target industries will follow. The priorities that cities attach to these alternatives are typically based on the assessment of the area's economic infrastructure. These observations lead to our next recommendation.

Recommendation Six: New Orleans should attach the highest priority to the "retain-and-grow" approach and the second-highest priority to attracting high-tech growth industries.

Organizational Phase

A major evaluation of New Orleans's prior development efforts (Bureau of Governmental Research, 2004) highlighted several problems. First, the city's prior efforts lacked an overall strategy and comprehensive development plan and relied instead on an ad hoc approach that created the perception that decisions were based on political rather than economic considerations. Second, there was no single individual or organization in charge of decisionmaking. Instead, decisionmaking was spread among multiple organizations, with the result that there were no clear priorities among alternatives. Third, no cost-benefit evaluation was conducted of individual projects or investments, so there was no way to determine whether they had met their objectives. Fourth, no attempt was made to assess the cumulative effects of individual decisions and their contribution to overall economic development.

Although there is a diverse range of public and private organizations at the state, regional, and local levels involved in economic development activities, many of the same problems remain. Specifically, these different organizations appear to have different emphases and agendas, represent diverse interests, and perform a wide range of functions. Although, in sum, they represent a valuable combination of resources and expertise, their efforts are not coordinated, reflect different priorities, lack continuity of personnel, and fail to follow a common strategy or plan.

In combination, these observations lead to our next three recommendations, relating to the structure, roles, and funding of the proposed economic development agency. Based on the experience of other cities, we suggest that New Orleans's economic revitalization program should have the following functional features: a comprehensive plan that is followed by all involved, a clear division of responsibility (e.g., for policy, implementation, resources), continuity of leadership commensurate with the term of the effort, and coordination of the activities involved in implementing the plan.

Correspondingly, we make the following recommendation.

Recommendation Seven: New Orleans's development agency should be led by a public-private partnership that consists of the following three components:

- A governing council comprised of leaders of the relevant public- and private-sector stakeholder organizations; the governing council should have direct responsibility for setting policy for the development effort and raising funds for that effort.
- A professional executive staff led by the director of economic development, who would be appointed by the development council for a five-year term and would report to the development council; the executive director would be responsible for implementing the plan and ensuring coordination of the development program's activities.
- A resource group drawn from the various organizations and individuals currently involved in development that would provide needed resources and expertise; the resource group would carry out specific tasks at the direction of the executive director.

The executive director's ability to ensure cooperation from the organizations in the resource group and to coordinate the development agency's activities would be enhanced by requiring him or her to specifically approve expenditures of funds that are directly related to the economic development activities.

Recommendation Eight: The funding for New Orleans's public-private economic revitalization partnership should be provided by both the public and private sectors.

Assuming that the costs of operating the agency will approximate the costs of running the city's Office of Recovery and Development Administration, the operating costs of the New Orleans economic revitalization partnership will be between $2 million and $3 million annually. Currently, the city is spending between $900,000 and $1.2 million per year on economic development activities. In addition, the city's economic development fund raised $5 million in fiscal year 2007 and is projected to raise more than $8 million in 2008. A significant share of the proceeds from this fund should be dedicated to the new agency, with a like amount provided by the private sector.

Recommendation Nine: The development council, in conjunction with the development executive, should consider appointing task forces that would be responsible for developing plans and implementation strategies for each of the target industry clusters.

Implementation Phase

Other cities place considerable importance on key implementation issues before embarking on their efforts. The criticism of New Orleans's prior efforts for failing to evaluate its programs' effectiveness, either individually or collectively, underscores the importance of such implementation issues. There are three specific questions that need to be addressed in the implementation phase. First, should the development effort concentrate on economic development at the city or the regional level? Second, what criteria should be used in setting priorities among the multiple goals of the effort? Finally, how will the effort measure progress, and will its plans be revised accordingly?

Most cities have adopted a regional approach for their development programs in the belief that their major competitors were other regions rather than other communities within their own region. Indeed, when considering a new location, prospective firms are likely to focus initially on the metropolitan area as a whole and only later consider in which specific communities they might choose to place their facilities. However, some of our interviewees in New Orleans told us that there was often competition between Orleans and the surrounding parishes and that prospective employers were sometimes drawn to the state by the attractions of New Orleans but ended up locating elsewhere in Louisiana. Our review of the evidence suggests that both the city and the region would both benefit in the form of increased employment opportunities and business for suppliers from growth and the entry of new firms, whether that occurs in the city or surrounding areas. This leads to our next recommendation.

Recommendation Ten: The focus of New Orleans's development efforts should be the metropolitan region rather than the city exclusively.

A successful development effort takes time and involves a wide range of activities. The list of improvements to the city's economic infrastructure discussed earlier, for example, included changes along several different dimensions, each of which will require a range of different actions. Moreover, the resources available to make those changes are limited. Correspondingly, a central issue for the development effort will involve setting priorities among competing goals. Thus, the development effort will need to determine what criteria to use to set those priorities. A variety of different criteria might be used. What is of paramount importance, however, is that criteria be chosen, that they be made public, and that their application be transparent. Following this guideline will go a long way to gaining credibility for the effort and gaining public acceptance for its actions. This leads to our next recommendation.

Recommendation Eleven: The development effort should set clear criteria for determining priorities among competing goals and actions, make those criteria public, and apply them in a transparent way. In particular, it should place highest priority on "the deal breakers"—that is, the actions and goals that are most critical to the success of the effort.

The final element of the implementation phase is to establish a means to assess the progress of the development effort. As we have noted, one of the criticisms of the city's past development program was its failure to assess the success of its efforts, either individually or collectively. As the experiences of other cities demonstrate, remedying this problem requires benchmarks against which to assess the progress and effects of a development program. This involves translating priorities into goals, incorporating those goals into schedules, and developing metrics to measure performance in meeting those goals. Based on the progress made in achieving these goals, modifications can be made, as needed, to the implementation plan. This leads to our final recommendation.

Recommendation Twelve: The development plan should include a process for assessing progress. This will require setting goals, establishing schedules, and developing metrics for determining the progress and success of its efforts.

Acknowledgments

We are particularly grateful to George Wentz and Arthur Pulitzer of the Horizon Initiative, which sponsored this work, for their support and guidance throughout the course of the research. We would also like to thank the many residents of New Orleans whom we interviewed for this research for their willingness to talk with us and provide their perspectives on the various challenges that the city faces.

This report also benefited from reviews by Robin Barnes, Robert Lang, and Henry Willis, whose comments improved both its form and its substance. Special thanks are also due to Robert Tannen and Jeanne Nathan for their advice and suggestions.

Finally, this report would not have been possible without the assistance provided by several RAND colleagues. Jeremiah Goulka and Sarah Cotton conducted most of the interviews for the research and made substantive and tactical suggestions along the way. George Penick, Michael Toman, Debra Knopman, Jack Riley, and Bernie Rostker provided advice, support, and practical assistance. Thanks are also due to Lauren Skrabala and Stacie McKee, respectively, for editing and guiding this report through the publication process. Finally, I would like to thank Isabel Sardou for her research assistance.

CHAPTER ONE

Introduction

The collapse of the levees and the flooding of New Orleans in the aftermath of Hurricane Katrina caused enormous damage and disruption to the city, its people, and its economy. Recovering from a disaster of this magnitude poses a major challenge to the city, the state, and the nation. The complexity of this challenge is compounded by the fact that New Orleans's population and economy had been lagging for more than a decade before Katrina.[1] Although recovery is under way, it has been proceeding more slowly than the city, its residents, and public and private leaders had hoped.[2]

In response to this situation, the Horizon Initiative, a private-sector organization, was formed in 2006 to expedite the city's recovery, with a particular focus on economic recovery. The goals of the Horizon Initiative are several. First, its overarching objective is not just to restore New Orleans's economy to the status quo ante but, rather, to reverse longer-term economic trends and revitalize the city's economy. Second, in the belief that this will require the efforts of both the public and private sectors, Horizon is working with the New Orleans City Council to develop an organizational and strategic program to achieve this end.[3] Finally, because economic recovery involves not just the city but also the greater New Orleans region, the Horizon Initiative is seeking to work with regional economic development groups (such as Greater New Orleans, Inc.) to advance its agenda.

As a result, the Horizon Initiative contracted with the RAND Gulf States Policy Institute (RGSPI) to provide recommendations as to the most effective organizational and strategic approaches to revitalizing the city's economy. This report is designed to address three questions relevant to organization and mission:

- What strategies should the effort employ?
- How should the effort be organized?
- How should priorities be assigned, and how should the approach be implemented?

---

1 Between 1990 and August 2005, prior to the storm, the population of the New Orleans metropolitan area grew 4.3 percent (compared with a national average of 19.2 percent), and total employment grew 7.9 percent (compared with a national average of 22.9 percent). The city itself has been losing population steadily since the 1970s. The city's total population of 600,000 in 1970 had declined to 485,000 in 2000 and to 455,000 in 2005 prior to Katrina. The city's current population has been estimated at around 250,000 (U.S. Census Bureau, undated[a], undated[b], undated[c], undated[d]; U.S. Department of Labor, undated).

2 See, for example, University of New Orleans (2007) and Liu and Plyer (2007).

3 Horizon presented its recommendations to the New Orleans City Council (together with the recommendations of Ed Blakely, head of the city's Office of Recovery and Development Administration) in November 2007.

RGSPI was asked to identify the best practices that cities have used to foster economic development (in terms of both how they have organized their efforts and the approaches they have used); to describe how these practices might be applied to New Orleans; and to recommend actions that might be needed to facilitate the application of those practices in New Orleans.

The purpose of this research is to provide an organizational architecture for New Orleans's economic revitalization, not a detailed redevelopment plan. By *architecture*, we mean the overall structure and mission of the effort, not a detailed blueprint. Although the content and approach of a specific economic plan or strategy are essential for putting it into action, creating that plan is beyond the scope of this report.

Methods

This analysis is based on three sources of information. First, using materials available on the Internet, we reviewed the development efforts and organizations in 17 cities.[4] This review enabled us to identify the key components of these areas' development approaches; the various strategies they pursued, including the rationale for these strategies; the alternative organizational arrangements employed in the different areas; and the various steps that were taken to implement these approaches. Second, we reviewed a wide range of published materials on the pre- and post-Katrina status of the New Orleans economy. These materials included both pre- and post-Katrina development plans, evaluations of the New Orleans economy and its strengths and weaknesses, assessments of the damage Katrina did to the economy and the status of the recovery to date, and reviews of past economic development initiatives and efforts in New Orleans. Finally, we interviewed a number of individuals in New Orleans, including public officials and representatives of the private sector. The selection of interviewees was made in consultation with the Horizon Initiative and public officials.[5]

In discussing the revitalization of New Orleans, this report draws a distinction between economic development and urban redevelopment programs. Economic development programs are designed to stimulate the economy by increasing the growth of employment and income in a city or region. Urban redevelopment programs, on the other hand, are designed to improve the physical, social, and economic conditions of neighborhoods within a city. Although both sets of efforts are important (and, as we discuss later, are related, since urban redevelopment is often an important component of increasing an area's attraction to new employers and residents), they have distinct goals and methods. Indeed, given the pre-Katrina economic trends mentioned previously, rebuilding the city absent growth in employment and income levels (two central goals of economic redevelopment) will result in a much smaller and less dynamic, if more habitable, city. This distinction is important because several of the communities we examined had efforts under way in both areas. Moreover, much of the planning effort that has been undertaken in New Orleans since Katrina has focused on rebuilding the city, including plans produced by the Bring New Orleans Back Commission, the Urban Land Institute, the Lambert Commission, and, most recently, the Unified New Orleans Plan. Although several

4 The organizations examined are listed in Appendix A.

5 The individuals interviewed are listed in Appendix B.

of these efforts have examined (and made recommendations) pertaining to economic development, their primary focus has been on the rebuilding of New Orleans's neighborhoods. This report focuses on economic development rather than urban redevelopment.

Finally, it is important to note that the process of successful economic development is complex. It is influenced not only by the strengths and weaknesses of the local economy and how they compare with those of other regions but also by national economic trends. Although the recommendations provided here are consistent with the experiences of successful economic development programs in other regions, they do not guarantee a successful economic development program for New Orleans. New Orleans must also be aware of national economic trends and developments, be responsive to opportunities those trends may present, and be willing to make adjustments as circumstances warrant. Finally, as we note repeatedly, the city must be open to the desires of the community.

Organization of This Report

The remainder of this report consists of three chapters. Chapter Two presents a review of economic development efforts in other cities, including the key lessons they offer for New Orleans. Chapter Three provides a review of some key features of the New Orleans economy, the local and statewide organizations involved in economic development, and critiques of the city's prior development efforts as a prelude to applying the lessons drawn from other cities. Chapter Four presents our recommendations.

CHAPTER TWO

Review of Economic Development Efforts

Given the importance of a healthy economy to the welfare of a city's residents and municipal revenues and services, most large urban communities devote a combination of public and private resources to promoting their areas' economies. Indeed, in the face of a changing global economy and increasing competition for business, many areas have boosted their economic development functions in an effort to support the growth of local firms and attract new firms—especially high-technology firms—to their regions.

This chapter examines the economic development activities of 17 urban regions to draw lessons from their experiences that might be applied to New Orleans.[1] Our purpose here is not to present detailed reviews of the specific activities of these cities or their development histories, since the applicability of one city's practices to another's will depend on specific circumstances and context. The comparison cities were not chosen based on a formal statistical design, and, given both the purpose of this study and the resources available, we do not attempt to construct a systematic mapping from city characteristics to successful development activities. Rather, we are interested in the general lessons we can draw for New Orleans from the design, organizational structure, and implementation approaches of the various cities. We can then use these general lessons, along with information particular to New Orleans, to develop a recommended organizational architecture for that city.

This chapter begins by identifying the areas we examined and the rationale for their selection. It then describes the various motivations behind these efforts; the key components of these development activities, including the varying strategies employed and how these efforts are organized; and, finally, some important aspects of how the programs have been implemented.

Description of Comparison Cities

Several criteria were used to select the comparison cities. First, we chose to focus primarily on cities that, like New Orleans, are located in mid- to large-sized metropolitan areas. Second, the Horizon Initiative had already been in contact with economic development agencies in selected cities and asked that these cities (e.g., Miami, Orlando, San Antonio) be included

[1] At the suggestion of one of the reviewers of this report, we also looked at the experiences of two other metropolitan regions that share New Orleans's location on a major river: Louisville, Kentucky, and Memphis, Tennessee. In general, the economic development programs in these areas operate similarly to those in the other areas examined. The one notable exception is Louisville's High Impact Program, which targets high-growth companies for special development assistance. See Greater Louisville Inc. (2007) for more information. For a description of Memphis's economic development program, see Memphis and Shelby County Office of Economic Development (undated).

in these comparisons. Third, we sought cities that shared some similarities with New Orleans, either in terms of their location (e.g., the South)[2] or the importance of key industries to their industrial structure (e.g., tourism, in Asheville, Charleston, Orlando, and Savannah; maritime shipping, in Charleston, Virginia Beach–Norfolk–Newport News, Miami, Savannah, and San Diego). Fourth, we selected four cities (Atlanta, Birmingham, Houston, and Jackson) that are located in close geographic proximity to New Orleans. Finally, we included cities with a range of different population characteristics and employment and income growth experiences.

The specific cities included in this analysis (together with data for New Orleans, shown for comparative purposes) are listed in Table 2.1. The table groups cities in terms of their income level and income growth (relative to the national average) for the period 1990 to 2005 and includes information on total population and employment growth over this period.[3] As these data indicate, five of these areas (Asheville, Jackson, Nashville, Orlando, and Winston-Salem) had lower income levels and slower income growth than did the country as a whole, though three areas (Asheville, Nashville, and Orlando) experienced faster-than-average population and employment growth. Two areas (Atlanta and Houston) had higher-than-average income levels (as well as more rapid population and employment growth) but slower income growth. Six areas (Albuquerque, Birmingham, Charleston, Miami, San Antonio, and Savannah) had lower income levels but faster income growth than the nation had. In addition, all of the areas in this category experienced faster population and employment growth than did the nation as a whole. New Orleans also falls into this category, but its population and employment growth lagged those of the nation.[4] The final category (higher income levels and faster income growth) included four areas (Jacksonville, Richmond, San Diego, and Virginia Beach–Norfolk–Newport News). Jacksonville is the only city in this group whose total population and employment exceeded the national average over the period. The specific organizations examined in these cities are listed in Appendix A.

Our objective in selecting these cities was not to rank their economic development programs (and thus look for a single best model), but rather to identify the most promising practices used across a range of cities to provide lessons for New Orleans in the different phases or components of an economic development program.

Characteristics of Economic Development Programs

Judging from the descriptions of their missions and histories, development efforts in all of these cities were motivated by a common set of factors: a recognition of the need to actively promote their areas as attractive places to do business in light of shifts in the structure of the U.S. economy and the increasing competition they face from other cities (both in the United States and abroad). In addition, although the emphasis placed on capturing firms involved in emerging high-tech industries varied, all contain programs designed to encourage emerging high-tech firms to locate in their areas.

---

[2] Fifteen of the 17 areas cited in this analysis are in the South.

[3] The end data for the comparison of trend data terminate in August 2005 for New Orleans to exclude the effects of Katrina.

[4] Indeed, New Orleans's population and employment growth (4.3 and 7.9 percent, respectively) not only lagged behind the national average on these dimensions but also lagged behind all 17 cities used in the comparison.

Table 2.1
Characteristics of Sampled Cities

	Income		% Growth Rate (1990–2005)	
Category and City	**Average (2005) ($)**	**% Growth (1990–2005)**	**Employment**	**Population**
U.S. average	46,242	53.9	22.9	19.2
Lower income, slower growth				
Asheville, North Carolina	38,275	51.3	26.3	27.5
Jackson, Mississippi	39,515	49.9	18.2	16.9
Nashville, Tennessee	45,543	50.7	34.0	35.7
Orlando, Florida	44,453	47.4	50.5	57.8
Winston-Salem, North Carolina	42,444	NA	15.8	24.1
Higher income, slower growth				
Atlanta, Georgia	54,066	51.8	50.7	60.2
Houston, Texas	46,705	48.4	29.4	40.2
Lower income, faster growth				
Albuquerque, New Mexico	43,070	57.7	29.3	33.1
Birmingham, Alabama	43,802	64.6	17.1	13.9
Charleston, South Carolina	45,269	61.3	23.8	17.4
Miami, Florida	43,091	60.1	33.8	33.7
San Antonio, Texas	43,263	66.1	38.9	34.2
Savannah, Georgia	42,444	57.0	38.8	21.6
New Orleans, Louisiana	39,879	63.3	7.9	4.3
Higher income, faster growth				
Jacksonville, Florida	47,323	60.3	32.3	34.9
Richmond, Virginia	52,524	56.8	22.1	23.9
San Diego, California	56,335	60.9	33.8	17.4
Virginia Beach–Norfolk–Newport News, Virginia	51,077	66.0	18.6	13.7

SOURCES: U.S. Census Bureau (undated[a], undated[b], undated[c], undated[d]) and U.S. Department of Labor (undated).

In addition, several of these areas had or were in the process of revamping their economic development programs for specific local reasons. In all of these areas, the economic development function had been spread across a range of public (city, county, or, in some cases, state) development agencies, as well as across various private organizations, such as chambers of commerce; tourism, convention, and visitor bureaus; and industry-specific organizations. As a result, many had become convinced that these various organizations were working at cross purposes, lacked a common development strategy, failed to coordinate their efforts, and were potentially transmitting competing messages. In addition, several areas were concerned that their national image was outdated or inappropriate for the kinds of industries and development activities they were seeking to attract. Finally, Charleston, which had been devastated by Hurricane Hugo in 1996, was involved in a major restructuring of its economic development program in conjunction with its recovery program.

Although, as might be expected, the specific development strategies and activities in these areas differed, our review suggests that the most impressive efforts were made by areas that incorporated three different components in their development programs: a strategy component designed to identify a desired future economic structure for the community (typically referred to as a statement of their vision for the future), a strategy for realizing that vision, and a set of target industries and activities for implementing the strategy; an organizational component designed to identify, recruit, and manage the human, organizational, and institutional resources needed to select and execute that strategy; and an implementation component designed to set priorities, establish a schedule, develop metrics for evaluating progress, and subsequently revise the implementation plan. We provide a more detailed description of these different components and the lessons they suggest for New Orleans in the remainder of this chapter.

Design Component

The objective of the design component is to produce an overall economic development plan. This plan typically contains several elements, including a statement of the vision for the future, a specific approach or strategy to achieve that vision, the identification of specific target industries or industry clusters, and a list of specific improvements to the local economic infrastructure (physical, human, and institutional capital) needed to attract and promote the growth of the target industries.

The vision statement describes the nature of the future economy that the development plan and the area's residents are trying to achieve. This vision is typically expressed in general terms, e.g., focusing on industries that promote balanced growth, are clean and environmentally sensitive, that pay higher wages and employ higher-skilled workers (Charleston Regional Development Alliance, undated), or a high-performance economy that creates greater economic opportunities for residents while enhancing the quality of life or providing economic opportunities (City of San Antonio, undated). These vision statements are expressions of a general philosophy or the guiding principles for future development, rather than detailed blueprints for the areas' development efforts.

Consequently, the process through which the vision statement is developed may be more important that its specific content. Typically, this process entails consultations with a broad range of stakeholders. For example, it often includes meetings with members of the business community (both heads of specific companies and members of business alliances, such as chambers of commerce), public officials (both elected officials and heads of local economic development and planning agencies), leaders of the academic and research communities, civic leaders, and the public at large (either in public meetings or through surveys). This process not only publicizes the economic development effort by soliciting the opinions of a diverse set of stakeholders, it potentially vests those stakeholders in the effort, securing their support. A broad outreach effort may be particularly important in New Orleans, where the range of stakeholders is broad, and prior experience (e.g., the post-Katrina neighborhood-planning process) suggests that, although difficult to achieve, consensus on such planning is critical to its public acceptance and success.

The next component of this phase is choosing an overall approach or strategy for the development effort. As identified in Table 2.2, there are three basic approaches that are used, each with a different set of targets, selling points, and tactics. The first approach, often referred to as the "buffalo hunt," is to recruit firms to relocate to the area (see MDC, 2002). This approach, the traditional choice of many Southern cities, has been used to recruit manufacturing firms—originally, textile and apparel firms, but more recently, automobile manufacturers (especially foreign manufacturers)—and their suppliers to the South. The principal tactics used with this approach are active marketing efforts, including marketing visits to firms and industry conventions, advertisements in trade and industry journals, and invitations to prospective firms to visit the area. The key selling points are the area's low costs (land, labor, and utilities), tax rebates, and a business-friendly local government and community.

A second approach focuses less on recruiting new firms than on retaining and promoting the growth of firms already located in the community. The major selling points of this approach are the area's economic infrastructure, especially those features of the infrastructure that are most important to the targeted firms; an adequate supply of qualified labor (often provided by training and education programs targeted to the firm' needs); and business-friendly government policies concerning such issues as permitting and zoning and tax incentives. The principal tactics used in this approach focus on internal improvements to the area's economic infrastructure that support the operations and growth of the targeted firms.

A third approach, and one that is being increasingly used by many cities, is to attract and sponsor emerging firms—particularly those in the high-tech sector. Since many, if not most, high-tech firms are dependent on a highly skilled labor force, a key determinant of their location decisions is the availability of highly skilled workers. Thus, these firms are drawn to areas that can attract relatively footloose, highly educated, and skilled workers. Such areas provide a high quality of life with natural amenities, good schools, and abundant arts and entertainment opportunities. In addition, high-tech firms are also attracted to areas with good universities and research capabilities, which also attract highly skilled labor. Thus, in contrast to the traditional buffalo-hunt approach, which is targeted to firms, this approach targets creative-sector workers on the assumption that, if the area can attract such workers, the high-tech firms will follow (see Florida, 2002, and DADCO Consulting Services, 2002).

Table 2.2
Three Strategic Approaches to Economic Development

Strategy	Goal	Selling Points	Approach
Traditional buffalo hunt	Relocate existing firms	Low costs Tax incentives Business friendly environment	Active marketing
Build on existing structure	Retain and grow	Infrastructure Available labor Business-friendly policies	Internal improvements
Growth industries	Attract and grow new-economy firms	Research capability Quality of life	Attract skilled labor to entice firms

Most of the communities we examined use a blend of these approaches but tend to emphasize one in particular. Their choice is typically based on several factors, including the strengths and weaknesses of the area's economic infrastructure (i.e., its current industry structure and its physical, human, and institutional assets), its comparative advantages vis-à-vis its principal competitors, and its ability to adjust to the demands of a changing economy. In addition, an important, if sometimes overlooked, aspect is the city's image both as a place to live and as a place to do business. A final factor that can influence this decision is whether the objective of the strategy is simply to promote economic growth (e.g., an increase in employment) or to restructure the local economy and its future prospects (e.g., economic development). This distinction may be particularly relevant to New Orleans to the extent that the city's goal is not simply to restore the local economy to its status pre-Katrina but to revitalize its economy.

The choice of a development approach is often made in conjunction with the decision to target specific industries. This decision is determined after an analysis of the area's current industrial structure, since virtually all development programs assume that the area's future economic growth will build on its current economic strengths. Three distinctions are often drawn in this analysis. The first is the determination of the relative importance (in terms of both employment and output) of the area's industries to its economy. The second is the distinction among industries as to whether they are primarily directed at external or internal markets. Underlying this second distinction is the premise that export industries are the principal drivers of local economic growth.[5] Finally, major export industries are then distinguished in terms of their future growth prospects. Those industries with reasonable growth prospects are generally designated as target industries.

There are, however, two common exceptions to this pattern. The first involves industries with limited growth prospects for which the local development group believes the area has a clear comparative advantage. Thus, the local area can offer persuasive inducement for such firms to relocate to their communities and capture a disproportionate share of future industry growth. The prime example of this is Southern cities' pursuit of automobile manufacturers. A second exception involves the decision of local economic development authorities to target emergent growth industries in which neither they nor other areas currently have a significant concentration of employment but the benefit to which the local area has other characteristics (e.g., a highly educated workforce, a strong university and research capability), offering firms in the industry a clear competitive advantage.

Although there is no golden rule governing the number of target industries that are selected, typically, no more than half a dozen or so are selected. This number is sufficient to offer a variety of targets, but not so high that development efforts are diluted by the pursuit of too many industries.

This description refers to industries, but, in fact, development strategies are typically focused less on specific industries than on clusters of industries. Industry clusters are defined as firms in the same or related industries that use similar production technologies, have similar production inputs (such as labor, materials, utilities, and services), and provide products to similar or related markets. These firms may operate cooperatively (as in the case of primary producers and their subcontractors) or competitively. The importance of the cluster approach is based on the fact that firms in related industries benefit from agglomeration economies by

[5] Note that tourism is viewed as an export industry because its market is comprised principally of outsiders who bring their dollars into the local economy.

sharing access to common inputs. The cluster concept bears a close resemblance to the "centers of excellence" concept with which the Horizon Initiative is already familiar.[6]

An important element of cluster analysis is the attention it draws to clusters of firms and the different sets of resources on which they depend. In fact, there are three components to an industrial cluster, each of which will affect an area's competitive advantage for attracting the cluster. The first is the export-based industries that constitute the core of the cluster. The second is the constellation of firms and industries that supply inputs and services (e.g., financing, advertising) to support the operation of the export-based industries. The final component includes the area's assets (e.g., training programs, academic institutions, research facilities, transportation facilities, labor exchanges), as well as the capital, financing, and other services that help ensure the successful operation of the core industries.

The relationships among these different factors provide a basis for the final element of the strategy phase: the identification of improvements to the local area's economic infrastructure and other characteristics that promote the success of firms within the cluster. These improvements can be quite broad, including upgrading the area's physical infrastructure (such as expanding and modernizing an area's port facilities; improving connectivity among rail, ship, air, and land transportation networks; alleviating highway congestion), its human capital (e.g., introducing new training programs designed to raise the skill level and supply of needed labor, improving the level of education that the community provides), providing available land for plant expansion, upgrading utilities (water, sewer, electricity, and internet connectivity), expanding research capabilities, increasing access to financial capital, general improvements to such public services as education and health care, and making local government operations more amenable to business operations (e.g., streamlining permitting and zoning processes, adopting a more customer-oriented approach to providing public services). In particular, these areas of emphasis include not just efforts to identify and encourage industrial clusters per se; they also include, among other actions, improvements to local governance and upgrading of the skill of the local workforce (human capital). Such improvements not only make an area more conducive to successful business operations, they can also improve the area's general image as a nice place to live and work.

Organizational Phase

An effective economic development program depends just as much on how the effort is organized and administered as on the strategies adopted. The major organizational challenge to such efforts is not that there are too few entities involved with economic development, but rather too many. As cities have come to recognize the importance of economic development programs to their economic vitality, the number and range of such organizations has multiplied. Within the public sector, for example, there are agencies at the state, regional, and local

---

6 Although clusters are usually defined in terms of specific industries, they can also be identified in terms of the labor pools from which they draw local workers with skills. This approach focuses on the supply of labor in the local economy as the most important of the community's assets. This approach may be particularly salient for attracting the new high-growth, high-technology industries that are the focus of the third strategy discussed here. Although this report uses the traditional approach of defining clusters in terms of specific industries, New Orleans may also want to consider a separate analysis based on workforce clusters and compare the results of these two approaches in terms of their implications for the city's economic development.

levels that are involved in economic development activities; indeed, there are often multiple organizations at each of these levels. These organizations have a range of functions, roles, programs, and resources (sometimes including dedicated funding from various governmental fees, such as hotel occupancy taxes). Although private-sector organizations involved in economic development activities are more likely to be focused on the regional or local area, they too have a diverse set of development targets, such as the metropolitan region in general; various communities within the region; specific industries, neighborhoods, and planning districts within the region; or ethnic communities and their business leaders.

The major challenge facing local development efforts is how to organize and coordinate these various players and their activities. Our analysis indicates that there are several different elements to solving this challenge. First, there needs to be a single overall plan or strategy to which the various organizations subscribe. Second, there should be a clear division of roles and functions among the various organizations involved in the effort. Third, there needs to be a clear division of responsibility for setting and implementing policies as well as assigning responsibilities for the specific activities that those policies identify. This division of functional responsibility typically includes three different functional tiers: an overall policymaking body that establishes and oversees the overall development effort, an executive staff that implements the strategy and coordinates the efforts of the various organizations involved in carrying out specific tasks, and an array of organizations that serve as resources to provide the specific expertise and services needed to implement those various activities.

These features of the organizational structure of the economic development effort appear to be particularly salient to New Orleans, given critiques of its prior development effort. As we will discuss in greater detail in Chapter Three, these critiques highlight difficulties that have arisen due to the absence of a comprehensive development plan, the failure to coordinate the many different agencies involved in economic development, and the absence of a lead agency to administer the various efforts (see Bureau of Governmental Research, 2004).

The composition of these different bodies typically varies according to their roles. The policymaking body often consists of public- and private-sector leaders, such as heads of major corporations, local government officials (both elected and appointed), heads of universities and research centers, and prominent civic leaders representing a diverse range of local communities. The status of this body provides the kind of legitimacy needed to ensure broad cooperation among the various organizations involved in the effort and the prestige needed to raise funding. The executive staff, on the other hand, is typically led by a full-time development professional who can provide the kind of expertise and continuity that is required to run the development effort. Finally, the diverse array of organizations and individuals in the third functional tier of this organizational structure may include many of the organizations that are currently working in this area and that have the specialized expertise and access to networks that are critical to contributing to the variety of activities involved in the development effort.

Increasingly, the policymaking and executive bodies are incorporated as a public-private partnership in recognition of the need for both the private and public sectors to cooperate and coordinate their activities as part of the total development effort. Indeed, the long-term nature of most economic development plans—typically at least five and often 10 years—requires a continuity of leadership that is unlikely among either elected officials or corporate executives.

Setting up this organization requires a concerted public-education campaign to familiarize the community and the various individuals and organizations involved with the goals, strategy, and importance of the economic development effort. The process of identifying the vision for the effort is an important component of this process. In addition, it requires a wide range of information about the community and its assets, broad media coverage, and meetings and briefings with various groups to ensure their cooperation.

Implementation Phase

The final component of these economic development programs includes four tasks: choosing a set of priorities among the various activities to be performed, setting up a schedule outlining short- and long-term goals, identifying a set of metrics to assess the progress in meeting these goals, and establishing a mechanism to review and revise the plan as experience warrants. As we discussed earlier, a comprehensive economic development plan includes a strategic approach, a set of target industries, and a list of desired improvements to the local economic infrastructure. Correspondingly, an effective implementation plan requires setting priorities for accomplishing these various activities. These priorities are typically based on three considerations: first, an assessment of the area's strengths and weaknesses, with the highest priority placed on those actions that are most critical to the success of the development effort;[7] second, an analysis of the area's comparative advantages and disadvantages relative to its competitors, which will help to identify those changes that will give the area a clear competitive advantage; and, third, an assessment of the feasibility and necessary resources to make the required improvements. The objective of this analysis is to identify those actions that are critical to the success of the effort, i.e., the potential deal breakers. These priorities are then manifested in a schedule for the development effort and form the basis for a series of short-, middle-, and longer-term goals. The deal breakers are given highest priority and thus made short-term goals, while those activities that require more resources or are less feasible in the short run are identified as longer-term goals.

Assessment of the progress of the development effort is the final element of the implementation phase. This element requires the identification of a series of metrics along which to evaluate progress toward the various goals, as well as a formal review process to determine what modifications may be necessary to the original plan.

Summary

Our review of the economic development efforts of 17 cities has revealed several important lessons.

- A successful economic development program requires a comprehensive design, an appropriate organization, and an effective implementation plan.
- A clear and comprehensive design starts with a vision of what the development authority and community are trying to accomplish, includes an appropriate plan or approach to achieve that vision that includes the identification of industries to target with the strategy, and a commitment to make the improvements to the area's economic infrastructure that will promote the growth of those target industries.

---

[7] The most critical status should be assigned to those activities that are deemed necessary conditions for success.

- Organizing for success is as important as the strategy used to achieve it. An effective organization requires coordination of the activities of the different parties involved in the effort with a clear division of roles and responsibilities among those parties.
- An effective division of labor is best served by establishing separate bodies to perform three key functions: a policy board to set policy, an executive board to carry out that policy and coordinate activities, and a resource group to provide the expertise to the effort as needed.

Finally, the implementation of the strategy requires setting priorities, goals, and metrics to evaluate the progress in meeting those goals.

CHAPTER THREE

Economic Development in New Orleans

The successful application of these lessons from other cities to New Orleans requires familiarity with New Orleans's pre- and post-Katrina economic situation and the organizational ecology of the city's economic development efforts.[1] Correspondingly, we reviewed selected studies of three different topics: pre- and post-Katrina studies of economic development plans for the city and the region, assessments of the damage done by the storm to the city's economic infrastructure and the progress it has made toward recovery, and reviews of the city's prior economic development efforts.[2] This chapter combines the findings from this review with the lessons from other cities' economic development programs to identify major issues for New Orleans's economic development efforts. This discussion is organized in terms of the three components of an effective economic development program, discussed in Chapter Two: design, organization, and implementation.

Design Component

As discussed, there are several elements to the design phase of an effective economic development program. These include a clear statement of the community's vision of what future economic development should look like, an assessment of the local area's economic infrastructure and its strengths and weaknesses, the identification of a set of key improvements to the area's economic infrastructure that will enhance its comparative economic advantage, and the choice of a specific development strategy or approach that includes the industrial sectors to be targeted.

Vision Statement

The first step in this phase is arriving at a common vision for the community's economic future. As noted earlier, the development of this vision is important not just as a means to set a common goal for the plan but, just as importantly, as a way to get the various stakeholders in the community involved in the planning for, and thus be vested in, the economic develop-

---

1 *Organizational ecology* refers to the configuration of organizations involved in economic development and the various functions they perform.

2 A variety of such studies is available. We reviewed the following studies, by topic area. Economic development plans: Bring New Orleans Back Commission (2006), Louisiana Economic Development Council (2003), MetroVision (2001), Regional Technology Strategies (2007), UNOP (2007), and Mt. Auburn Associates (2005). Damage done by Katrina and the current state of recovery efforts: Terrell and Bilbo (2006), Liu and Plyer (2007), and University of New Orleans (2007). Development progress: Bureau of Governmental Research (2004).

ment effort. This step may be particularly important in New Orleans, both because the city's prior economic development activities have been criticized as lacking an overall development plan (see Bureau of Governmental Research, 2004) and because, as the planning for rebuilding the city has demonstrated, reaching consensus among stakeholders about the future of the city appears to be particularly difficult.[3]

Our interviews indicated that the city contains a diverse range of stakeholders with different interests and perspectives on New Orleans's future. In addition to overcoming its history of racial fragmentation,[4] reaching consensus on the city's future will be complicated by different perspectives between the public and private sectors, between large and small businesses, among neighborhoods, and between those who focus on the city and those who approach redevelopment from a more regional perspective. If the city's economic development efforts are to be successful, efforts must be made to bridge these divides and reach a consensus on the direction and organization of the development effort. Correspondingly, the vision statement should incorporate the views of a diverse range of stakeholders. Although reaching consensus among these various groups may be difficult, it is important that a wide range of perspectives be solicited and heard.[5]

To facilitate this outcome, it is useful to draw a clear distinction between the city's economic development and rebuilding efforts. Although these two efforts are not unrelated, the experience of planning for rebuilding the city's neighborhoods has demonstrated that decision-making regarding which areas and neighborhoods to rebuild and when raised a host of contentious issues that are not central to rebuilding the city's economy. Rather than revisit these issues and the conflicts they spawned, a vision statement should focus on such key economic development issues as the types of jobs and industries that should dominate the city's future economy.

Moreover, as we noted earlier, rebuilding the city's neighborhoods without reinvigorating its economy will result in a much smaller and less dynamic city. Indeed, to the extent that the city's (and the region's) economic development effort is designed not simply to restore the city's economy to its pre-Katrina status but to revitalize it for the future (e.g., economic development rather than basic economic growth) it will require not simply repairing the damage that Katrina did to the city and regional economies, but also solving the structural and economic problems that were limiting the city's economic prospects well before Katrina.

Assessment of Economic Infrastructure

Prior to Katrina, New Orleans's economy was based on four key industrial clusters: the maritime industry (shipping, shipbuilding, and ship repair), oil and gas (production and petrochemicals), hospitality (accommodations, food, and a host of tourism-related activities), and biomedical services (health services, medical research, and other areas). In addition, the city

---

[3] There have been at least four separate plans proposed for rebuilding New Orleans in the aftermath of Katrina. Each plan was announced with considerable fanfare but was then subject to considerable criticism by various stakeholders and subsequently dropped in whole or in part. The current plan (UNOP, 2007) appears to have received wider public support, in part because it was developed after obtaining input from a wide and diverse range of stakeholders.

[4] The majority of those we interviewed cited race as a major factor dividing the city, both in the past and today. Indeed, at least two respondents said that Katrina had exacerbated racial issues in the city.

[5] Judging from our interviews, it will not be difficult to get opinions of a wide and diverse range of individuals about the problems the city faces and what should be done about them.

had notable manufacturing activity in the food-processing and aerospace industries, as well as notable clusters of employment associated with its centers of research in medicine, biotechnology, and computer technology. The city also had a long tradition of arts-related employment in the entertainment industry, with a significant concentration of musicians and a growing film and television production industry.[6] Supporting many of these economic activities was a strong and diverse higher-education and research sector. Finally, a dynamic small-business sector accounted for 40 percent of the city's employment (BNOB, 2006, p. 7).

In addition, MetroVision's pre-Katrina report on cluster development in southeast Louisiana identified nine different industrial clusters (each containing a set of detailed industries) that could serve as economic drivers in the region (MetroVision, 2001; see also DADCO, 2002). It sorted those clusters into four different categories based on future growth prospects and the industries' current levels of maturity. The mature cluster, which identified industries with moderate growth prospects and a well-established base in the region, included oil and gas, maritime, and food and consumer products. The dynamic cluster, which identified industries with healthy growth prospects and a solid foundation in the region, included the medical and petrochemical industries. The emerging cluster, which identified industries with promising growth prospects and an initial foundation in the region, included the arts and entertainment, environmental technology, and information technology industries. The final (developing) cluster, which identified industries that could expect healthy growth and with an initial foundation in the region, included the life sciences.

In their review of significant university, federal, and private-sector research facilities along the Gulf Coast (defined as the four states along the Gulf of Mexico stretching from Florida to Louisiana), Tortorano and Tortorano (2006) identify nine major research clusters in the region.[7] The New Orleans area has at least one research facility in each of these clusters and significant concentrations in several, including federal shipbuilding and repairs, computer sciences, medicine and biotechnology, and earth and environmental sciences. As Tortorano and Tortorano point out, each of these clusters is expected to grow well above the national average in the future, and a solid research foundation provides the key element needed for local areas to capture that growth.

Although Katrina did substantial damage to the city's economy and its industries, both the maritime and oil and gas clusters have made solid recoveries, and the hospitality cluster appears to be on its way (Liu and Plyer, 2007; University of New Orleans, 2007). Moreover, New Orleans has a clear competitive advantage in several of its core clusters, given its location on the Mississippi River, its access to railroad networks, its proximity to oil and gas reserves in the Gulf of Mexico, and its numerous cultural and tourist attractions—and these advantages have not been altered by Katrina. On the other hand, the biomedical cluster suffered serious damage with the closure of the Louisiana State University Medical Center and several other hospitals and the loss of a significant portion of its health professionals, as well as serious damage to much of its research capabilities. Moreover, although higher-education and research centers have mostly reopened, they have not yet regained the capabilities that were critical

---

[6] New Orleans and Louisiana were collectively the third most frequent location for such production in the United States (BNOB, 2006).

[7] The research clusters include aerospace/defense, federal shipbuilding, geomatics, computational science, energy/oil and gas extraction, marine sciences, medicine/biotechnology, and earth and environmental sciences.

to their economic importance. Finally, the city's small-business sector was devastated by the storm (Terrell and Bilbo, 2006).

A key lesson from other cities is that effective development programs should start by building on an area's existing economic strengths or—in the case of New Orleans—restoring its key strengths. Correspondingly, a high priority for the city's development efforts should be to focus on restoring, where practicable, the city's preexisting advantages for its key industrial clusters and rebuilding those industries that were badly damaged by Katrina to their pre-Katrina prominence. This will require major capital investments in the infrastructure supporting several of these key clusters, as well as a change in the city's policies for dealing with the business sector (see "Internal Improvements: Infrastructure, Workforce, and Governance," later in this chapter). In each case, consideration would need to be given to the degree to which pre-Katrina advantages could be effectively recaptured by new investments and policies.

Although the need for many of these infrastructure improvements, as well as changes in the way the city interacts with its business community, predate Katrina, the storm exacerbated many of these and other weaknesses in the city's economy. These include an aging physical infrastructure (roads, sewers, and utilities) that needed major improvements before Katrina, a labor force that lacks needed skills, and a port system that had been losing shipping traffic to other U.S. ports before the storm.

Katrina also highlighted in a most dramatic fashion several other weaknesses in New Orleans's economic infrastructure. Perhaps the most significant example of the latter, though one over which the city lacks direct control, is the status of the levees and the flood protection they will provide in future storms. Uncertainty about the level of protection the levees will provide from future storms raises serious questions about the wisdom of investment in the city, not only among current and displaced residents but also among potential newcomers and businesses.

Two other major casualties from the storm over which the city has more direct control are the city's image and the quality of life it provides for its residents. Even before Katrina, politics in New Orleans and in Louisiana more generally suffered from a lack of transparency and a perceived tolerance for corruption.[8] But Katrina not only devastated New Orleans's physical and human capital; it also further damaged the city's image and reputation. As the experience of other cities' economic development efforts indicates, how the city is perceived externally (both as a place to live and as a place to do business) is critical to its ability to attract both businesses and workers. Careful thinking about what steps can be taken to change this image in the short term are likely to pay enormous dividends for the economic future of the Crescent City.

A second major casualty of Katrina was the quality of life New Orleans offers to its residents. Despite its many charms, New Orleans is now a difficult place to live for most of its residents. If the city is to encourage its displaced residents to return and to attract both new residents and businesses, it will need to improve the quality of life. This will require the city to address several major problems. Currently, New Orleans has the highest violent-crime rate of any city in the country. Improving the quality of life will require addressing this

[8] This point was underscored in our interviews and is also reflected in media coverage of the city after the storm. See, for example, Fruth (2007) and Thomas (2007). In commenting on the lack of civic involvement in the state and the city, Charles Mann, writing for *Fortune* magazine, quoted Shreveport political analyst Elliot Stonecipher: "Nobody in Louisiana knows what noblesse oblige is. New Orleans is a hotbed of civic apathy—the only city in the country where rich and powerful people don't have their fingers in everything" (Mann, 2006, p. 92).

crime rate so the city's residents feel safe and secure. In addition, the city needs to create effective schools that include day-care options. Although the new charter schools are a promising start, the city must strengthen its entire education system.

The city also needs to rebuild a health care system that was one of its strongest assets, not only as a center for health delivery but also for innovation, teaching, and research. This will be a challenge given the damage that was done to the biomedical infrastructure and the loss of a significant fraction of the cluster's professional personnel. Several of the city's major medical facilities, for example, were severely damaged or destroyed by the storm (including Charity Hospital) and not all have reopened. In addition, initial estimates suggested that more than half of the city's professional medical staff (doctors and nurses) left the area in the immediate aftermath of the storm (AngelouEconomics, 2007). Although many have subsequently returned, total employment in the health care sector in the New Orleans region was still down almost 40 percent at the end of 2006.

In addition, the city needs to ensure that a sufficient supply of affordable housing is available and to restore basic services to newly rebuilt and rebuilding neighborhoods. The quality of life of the city's residents depends on their ability to find suitable housing at an affordable price and to count on basic public and private services.

Internal Improvements: Infrastructure, Workforce, and Governance

Building on its strengths and addressing the weaknesses of its economic infrastructure will require the city's public and private sectors to make a number of improvements to that infrastructure. Indeed, it was common practice in many of the cities we examined to lay out a plan to address such internal improvements, whatever specific development strategy they adopted. Many of the improvements described here have already been identified in the Bring New Orleans Back Commission's economic development report (BNOB, 2006) and by the Unified New Orleans Plan (UNOP, 2007). The list presented here includes upgrading the city's physical infrastructure, developing its workforce base, changing government policies toward business (including special efforts to reinvigorate the city's small-business sector), and upgrading the city's image and the quality of life for its residents.

Infrastructure. Despite the port of New Orleans's natural advantage, it had been losing tonnage and value prior to Katrina. In part, this reflects the failure to invest in the port to expand its capacity and its capabilities. Several improvements have been identified elsewhere to develop the port's capabilities, including upgrading its ability to handle containerized shipping, relocation of a new cold-storage facility, and a new cruise terminal (see BNOB, 2006, and UNOP, 2007). Both the cold-storage facility and the cruise terminal have an added advantage in that they will also benefit the food-processing and tourist industries.

As noted earlier, New Orleans's medical facilities were a major asset that was badly damaged by the storm. They served as the major provider of health services to the region, as well as a center for medical research and an incubator for the biomedical industry. Rebuilding the medical industry's infrastructure for the provision of medical services should be a high priority for the city—particularly given the expected growth of this sector in the next decade as the region's and the nation's populations age. Consideration also should be given to the prospects for rebuilding the medical and biomedical research infrastructure as a focus for cluster development, given the expected growth in this sector in the future. Key infrastructure improvements include the Louisiana State University/Veterans Affairs medical center, the completion of the Cancer Research Center, and the biotechnology business innovation center. The rebuilding of the biomedical cluster will also need to address the loss of the professional workforce in the sector.

As the Unified New Orleans Plan report (2007) indicates, several capital projects have been considered to upgrade the city's tourism facilities, including an expansion of the airport, the Canal Street and downtown revitalization program, and the previously mentioned addition of a cruise terminal at the port. The expansion of the airport might also include the addition of cargo facilities that would improve the city's trade capabilities, especially with Central America if the proposed free-trade agreement with Central America is finalized.

Finally, as noted earlier, the damage Katrina did to the levees and the city's already aging and deteriorating infrastructure will need to be addressed. Upgrading this infrastructure will, of course, affect the city generally rather than just providing benefits to specific industries. Repairing the levees is probably the single most important capital investment needed, but the city has no direct control over this decision.

Workforce. In the immediate aftermath of Katrina, the population of New Orleans was decimated and, although many of the city's residents have gradually returned, the total population is still well below its pre-Katrina level.[9] One by-product of this outflow has been a very tight labor market that has made it difficult for businesses in the city to find sufficient numbers of workers to fill their demand for both high- and low-skilled jobs. As a result, wage levels have climbed sharply since the storm.[10]

But problems with the New Orleans labor market predate Katrina and include both an overall shortage of workers and a shortage of workers with the skills needed by businesses. The city was, for example, losing residents, especially its highly skilled and well-educated population, for at least three decades prior to the storm. This problem has intensified since Katrina. Indeed, one of our interviewees indicated that 20 to 30 percent of the applicants to workforce development programs fail to qualify because they lack basic skills. Since a sufficient supply of quality workers is essential to a successful business environment, New Orleans needs to address its labor situation.

Since the quality of an area's workforce is rooted in the quality of its education system, improving the quality of New Orleans's schools is clearly needed, and steps are currently being taken to improve education in the city. But the payoff from this investment will not be realized in the short term. Thus, in addition to the long-term investment in its schools, the city needs workforce programs that will begin to improve the quality of its labor pool today. Better vocational training programs at the high school and community college level would be a good first step (including internships with employers), as would job training programs that are focused on the current and future labor needs of business.

Government Policies Toward Business. A common complaint among the interviewees was that the city's interaction with the business community needs to be improved. Specific suggestions included a more efficient and transparent permitting and zoning process, a one-

---

[9] Precise figures on the city's current population are hard to come by, but a recent report by the Brookings Institution and the Greater New Orleans Community Data Center (Liu and Plyer, 2007) estimates that, by August 2007, the city's population was about two-thirds of its pre-Katrina total. Other studies suggest that the city's repopulation will be a slow process (see McCarthy et al., 2006).

[10] A report by the University of New Orleans (2007), for example, estimates that, by the end of December 2006, the average weekly wage in the metropolitan region had climbed to 126 percent of its pre-Katrina level and, in many sectors, substantially higher. It is unclear what percentage of this increase represents real wage growth and what percentage is consumed by the rising costs of living in post-Katrina New Orleans.

stop shopping approach for businesses making inquiries and applications for licenses and permits, and a more service-oriented approach by government employees. Several interviewees, for example, complained that the city had never contacted them about their needs or whether the city could improve its service delivery systems. This situation stands in sharp contrast to the experience of other cities in which local governments make a determined effort to assist business by providing one-stop shopping as well as an advocate or case manager to help businesses negotiate with local government.

In addition to improving the city's interaction with the business community, the city and the development effort should also analyze Louisiana's development programs, including the multiple incentives that they provide for business development. These include business tax incentives, direct financial assistance, community development services, and the identification of key industries in which the state will concentrate its economic development activities.[11]

Dealing with the Small-Business Sector. As noted previously, the small-business sector played a vital role in the city's pre-Katrina economy, supplying about 40 percent of the city's jobs and providing entrepreneurial opportunities for the city's residents, especially its minority residents. However, this sector was devastated by Katrina (Terrell and Bilbo, 2006; University of New Orleans, 2007) and will need assistance to get back on its feet. Although there are a variety of ways to provide this assistance, the focus should be on the principal problems that small businesses are likely to face (e.g., technical assistance, access to capital, business incubators to try out new ideas). Each of these issues—but especially access to capital vis-à-vis creditworthiness—are significant institutional challenges. Moreover, because the small-business sector has traditionally been an entry point into the system for minority entrepreneurs, efforts should be made to link these efforts to the various interest groups within the city.

Upgrading the Quality of Life. In addition to working to improve the city's economic infrastructure, two other improvements are vital to the city's economic recovery. The first is to upgrade the quality of life of the city's residents. This will require the city to address a crime problem that has become notably more severe over the past two years and has drawn the attention not just of the city's residents but the national press as well, to restore (and improve) the city's medical and educational services, to provide an adequate supply of affordable housing, and to restore basic services to the city's rebuilding neighborhoods. Successfully addressing these problems will not only make living in the city more agreeable, it will also improve the city's image both to its own residents and to outsiders. As such, it will send a clear signal to those outside the city, including the city's many displaced residents and potential migrants that now is the time to consider returning or moving to New Orleans.

Rebranding New Orleans. The New Orleans brand suffered from a perception of corruption and cronyism even before Katrina. The difficulties of post-Katrina recovery and additional indictments of local officials have further tarnished the New Orleans image. Restoring the city's image and, thus, the New Orleans brand is not a sufficient condition for development alone, but it is nonetheless critical to the city's economic recovery. It is difficult to overestimate the importance of this improvement for the city's economic development. One of the clearest lessons to be learned from other cities is the emphasis they place on how outsiders view their cities, not just as a place to do business but also as a place to live. Indeed, many cities, particularly those that want to change the way they are viewed, attach such importance to their city's

---

[11] See Louisiana Economic Development Department (2007). In addition, the state's development council has issued a master plan for economic development (Louisiana Economic Development Council, 2003).

brand that they invest considerable sums in "rebranding" their city's image, including developing a new logo to represent the new image. Given the bad publicity that New Orleans had attracted even before the storm, the city's economic development effort may want to consider a formal rebranding effort.

Setting Priorities. This discussion is not meant to provide a definitive list of the improvements that New Orleans needs to consider, but, rather, to suggest how important it for the city to recognize what needs to be done in New Orleans and by New Orleans to create conditions that will make the city a more conducive place for its key industries to stay and grow and to attract new and displaced businesses and residents.

Since the list of improvements discussed here is long and it is unlikely that they can all be accomplished in the short term, a key decision the city faces is how to set priorities among the alternatives. Selecting priorities will require a more detailed examination of the city's current economic infrastructure than is possible here, but our analysis suggests some general considerations to bear in mind. First, even before Katrina, many of the economic trends in the city were moving in the wrong direction. Correspondingly, it is important for the city to have some quick and visible successes to indicate that the situation is turning around. We suggest that priority be placed on those improvements that are both highly visible and feasible to accomplish in the short term. Second, as we have pointed out, several of the improvements mentioned can affect multiple dimensions of the city's infrastructure and its industries simultaneously. This suggests that priority be placed on those improvements that will have the greatest "bang for the buck." Third, costs will obviously affect the feasibility of making these improvements.[12] Consequently, priority should be given to improvements for which private or public funding is readily available. Finally, the city's prior economic development activities were not based on a comprehensive plan. Rather, they appear to have been based on an ad hoc approach that has been criticized as serving specific stakeholders' interests rather than those of the city as a whole. To prevent that from recurring, the plan for making improvements and setting priorities among the alternatives should be transparent and derived from a comprehensive plan that can gain concurrence from a broad segment of the city's stakeholders. What is most important, however, is that the recognition that the city's economic recovery will require changes in how business has traditionally been done in New Orleans and that the primary responsibility for this change rests with the city, its residents, and its business community and not with outsiders.

Choice of Development Strategy

The final component of the design phase is the choice of a development approach or strategy together with the identification of the industries to target with that strategy. As discussed in the previous section, there are three basic approaches that might be used. The first is the traditional buffalo-hunt strategy that seeks to recruit new industries to the area. The second is the retain-and-grow approach that aims to strengthen the city's existing industrial base and create conditions that are conducive to those industries' growth. The third approach focuses on creating conditions that will promote and develop new high-tech, high-growth industries. Although most cities choose some combination of these approaches, they tend to emphasize one approach over the others. Their choice is typically made based on how suited their eco-

[12] The Bring New Orleans Back Commission's economic redevelopment plan, for example, estimated that almost $4.9 billion of direct federal aid would be required to repair the city's economy. This is in addition to state and local funding (BNOB, 2006).

nomic infrastructure is to the key selling points on which the different approaches depend. The buffalo hunt emphasizes low costs and a business-friendly image. The retain-and-grow approach emphasizes developing the infrastructure, labor, and government policies that support the growth of the city's existing industries. The high-tech, high-growth industry approach emphasizes building the area's education and research capabilities, as well as maintaining a quality of life that will attract highly skilled and highly educated workers.

In light of New Orleans's current situation, the primary strategy for its economic development should be the retain-and-grow approach, with the high-tech, high-growth strategy given secondary emphasis. Although there may be selective opportunities to recruit new industries over the longer term, the buffalo-hunt approach is ill suited to the city's current economic situation. The success of the retain and grow strategy will hinge on the city's ability to implement the infrastructure, labor, and government policies needed to promote the successful operation of the city's key industries and its small-business sector. In addition, restoring the city's research and educational capabilities and, most particularly, upgrading the quality of life of its residents offers promise for the high-tech, high-growth strategy, given New Orleans's pre-Katrina base in the research area (Tortorano and Tortorano, 2006).

In implementing this approach, the development plan should target three groups of industries. The first group (oil and gas, maritime, and tourism) contains the core industrial clusters. The second group (medical services, food processing, biotechnology, and the small-business sector) contains core industries that suffered major damage from Katrina that will have to be repaired for them to resume their prior prominence. This will, of course, require balancing the pros and cons in each case. The third group (arts and entertainment, energy technology, and information technology) are clusters that had a solid footing in the city before the storm and are potential growth sources for the future. Using these industries as the focus should help establish priorities for the development effort.

Organizational Component

The second component of an effective program is the choice of an appropriate organizational structure around which to develop and implement the strategy. Our discussion of this phase is based on reviews of the city's pre-Katrina experience and the current organizational ecology of the city's development programs. Based on these reviews and our examination of other cities' programs, we then draw lessons for New Orleans.

New Orleans's Prior Development Experience

In 2004, the Bureau of Governmental Research, a nonprofit think tank in New Orleans, published a study of economic development funding in the city during the five-year period from 1998 to 2002. Although its specific focus was on funding practices, the study included a critique of economic development activities in the city. This critique underscored several major problems with the economic development effort. First, it noted that the $1.1 billion that was spent was not based on an overall development strategy or plan but relied instead on an ad hoc approach. As the study report put it,

> The planning process in New Orleans stops short of the critical step of establishing priorities among competing goals, leaving unanswered fundamental questions concerning

> the allocation of resources. In addition, expenditures tend to be considered on an ad hoc basis, without evaluation of their relative value or the "opportunity costs" of a particular investment. As a result, public investments, particularly in private ventures, run the risk of being based on political, rather than economic, considerations. (Press release accompanying Bureau of Governmental Research, 2004)

The study further noted that the city's development activities suffered because there was no "investor" in charge. Instead, authority was divided among multiple organizations. As a result, there was no clear set of priorities. Instead, key decisions were driven by individual funding organizations' agendas. Further, no evaluation was conducted on whether projects met their promises, and no analysis was undertaken to assess the cumulative effects of the individual grants. Indeed, there was no cost-benefit assessment of the various development projects and how they might contribute to the city's economic development.

The interviews we conducted with individuals in New Orleans for this study confirmed these conclusions and indicated that this situation had not changed between the period of the study and Katrina or since Katrina. Correspondingly, we examined the various public and private organizations in the city and the state to determine what organizations were involved and what roles they play in economic development.

The Organizational Ecology of Economic Development

Currently, there is a diverse array of organizations involved in some way with economic development in New Orleans. These organizations operate at the state, regional, and local levels and focus on a wide range of activities and purposes. Figure 3.1 presents a schematic overview of these organizations, the geographic levels at which they operate, and the specific focuses of their activities.

The principal economic development agency at the state level is the Louisiana Department of Economic Development.[13] The department administers one set of programs, primarily for small businesses (e.g., loans and seed capital), through the Louisiana Economic Development Council, and a broader set of business incentives through the Louisiana Board of Commerce and Industry. In addition to these broadly targeted economic development programs, the lieutenant governor has established the Louisiana Cultural Economy Foundation to promote the state's culture, recreation, and tourism industry.[14] In addition, there are several statewide private business organizations that seek to promote specific business groups or industries, such as the Minority Business Council.

There are two sets of organizations that focus on economic development at the regional level: The first focuses primarily on economic development, the second on urban redevelop ment. The first set includes Greater New Orleans, Inc., the New Orleans Regional Chamber of Commerce, and the New Orleans Regional Loan Corporation. Greater New Orleans, Inc., is a public-private partnership that fosters economic development in the 10-parish greater New Orleans region by working to create jobs, market the parishes to companies seeking to expand or relocate, and to retain and grow existing businesses (see Greater New Orleans, Inc., 2007).

---

[13] A description of programs can be found at Louisiana Economic Development Department (2007).

[14] The Louisiana Department of Culture, Recreation and Tourism issued a strategic plan for Louisiana's arts and culture sector in July 2005 (see Mt. Auburn Associates, 2005).

Figure 3.1
Organizational Ecology of Economic Development Efforts

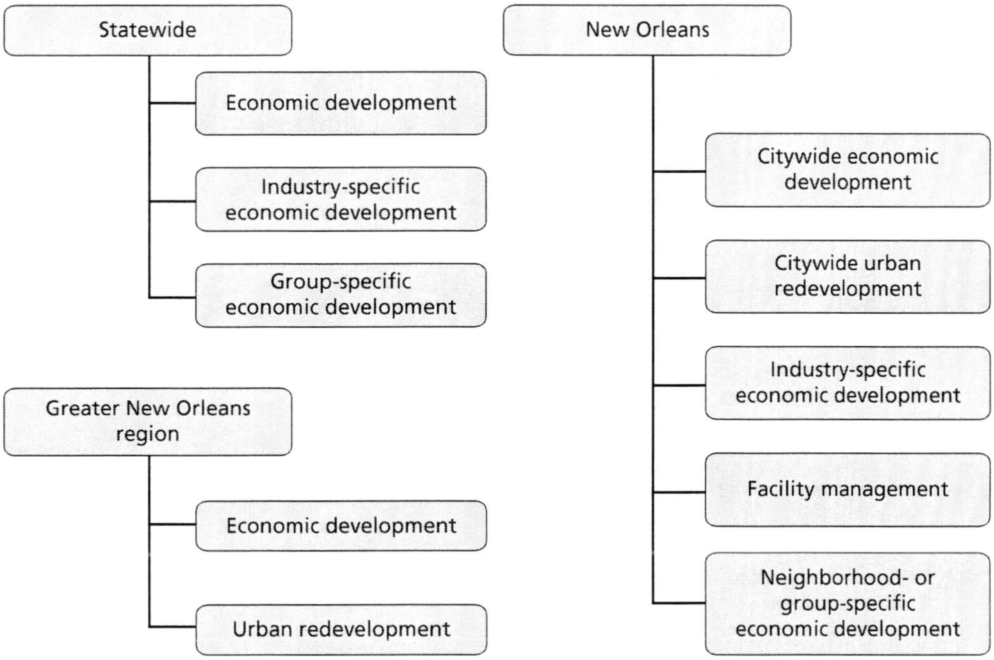

RAND *TR547-3.1*

The New Orleans Chamber of Commerce is a membership organization for companies in seven parishes in the greater New Orleans area that advocates for its members on public-policy issues affecting the business community (see New Orleans Chamber of Commerce, undated). The Regional Loan Corporation works with the U.S. Small Business Administration and the Economic Development Administration to provide loans to small businesses (see Regional Loan Corporation, 2006). The second set includes the New Orleans–Jefferson Renewal Community and the Regional Planning Commission. The first of these organizations is a cooperative arrangement among parishes to renew impoverished neighborhoods by offering tax incentives for businesses and residents through U.S. Housing and Urban Development's empowerment zone and renewal community programs. The Regional Planning Commission is a comprehensive planning agency that includes regional public- and private-sector leaders (see Regional Planning Commission, undated).

There is an even larger and more diverse set of organizations working on economic development issues at the local level. There are, for example, several agencies within the city government that work on some aspect of economic development, including the Office of Recovery and Development Administration (set up after Katrina to oversee the city's recovery efforts), the Office of Planning and Development (which controls a significant share of the city's economic development fund), the Finance Authority (which provides tax-exempt financing for public facilities, facilities owned by nonprofits, multifamily housing, and other forms of community and economic development), and the city council. There are also several separate private-sector chambers of commerce working with different segments of the business community, including the various underrepresented communities.[15]

The city's Office of Recovery and Development Administration has overall responsibility for rebuilding and urban redevelopment, but there are also other city agencies that are involved in redevelopment, including the New Orleans Redevelopment Authority,[16] the New Orleans City Planning Commission, and the Division of Housing and Neighborhood Development.[17]

In addition to those organizations working on citywide issues, there is also a host of organizations working on industry-specific development and neighborhood-specific redevelopment. The former includes the Naval Support Activity's New Orleans Advisory Task Force, the Industrial Development Board, the New Orleans Tourism Marketing Corporation, the New Orleans Metropolitan Convention and Visitors Bureau, the New Orleans Workforce Investment Board, the utilities' economic development funds, the Algiers Development District, the Downtown Development District, the Vieux Carre Commission, and the Historic District Landmark Commission.

Finally, a variety of boards administer various infrastructure facilities and are directly related to the city's economic development. These include the Ernest L. Morial New Orleans Exhibition Hall Authority (the convention center), the Board of Commissioners of the Port of New Orleans, the Louisiana Superdome and Exposition District Board, the Audubon Commission (the zoo and aquarium), the New Orleans Business and Industrial District regional business park, and the New Orleans Aviation Board (the airport). In addition, the city has several public-benefit corporations that administer activities that are commercial in nature and generate their own revenues.[18]

Implications

This organizational ecology represents a valuable combination of resources and talent, but, given the diverse nature of their activities and their specific missions, it is not surprising that coordination of their combined activities is problematic. In addition, since they receive their funding from a variety of different sources, including member dues and assessments, appropriations from state and local governments, dedicated tax levies and fees, and a variety of grants and contracts, they value their independence and have no obligation to pursue a common development strategy. Indeed, since they are typically formed to promote the specific interests of their members or the clientele they serve, those specific interests—rather than the broader community's interests—are their highest priority. In addition, each of these various organizations has its own board or director, which may change over time as newly elected officials name their own appointees, as terms of office expire, and as individuals move on and are replaced, and, consequently, their specific policies may change over time.

This situation is not unique to New Orleans. Indeed, it is one of the primary reasons that other cities have reorganized their economic development functions. The objective of these

[15] The New Orleans Chamber of Commerce merged with another private-sector business group, MetroVision, to form Greater New Orleans, Inc., in 2004.

[16] See New Orleans Redevelopment Authority (undated).

[17] See City of New Orleans (undated).

[18] These enterprise operations include the Upper Pontalba Building Restoration Corporation, the Rivergate Development Corporation, the Piazza d'Italia Development Corporation, the Municipal Yacht Harbor Corporation, the French Market Corporation, the Canal Street Redevelopment Corporation, and the Delgado-Albania Plantation Commission.

reorganizations is not to ignore or displace existing organizations, but to coordinate their efforts and work within an overall strategic plan. These types of organizations have several common features, including

- a common set of goals and a clear, strategic plan that all groups follow
- a division of responsibility or authority with a governing board that has wide representation of the various stakeholders in the development effort and a professional executive who implements the plan
- continuity of professional leadership for the executive authority
- coordination among the various groups and activities involved in carrying out the plan, including both the public and private sectors
- recognition of the importance of involving the public and private sectors in these efforts, typically through a public-private partnership.

In light of the experiences of other cities and the specific circumstances in New Orleans, the organization of economic development activities should consist of three tiers. The first tier would consist of a general governing council with both public- and private-sector membership that is responsible for setting policy (e.g., developing an overall strategy, selecting target industries, setting priorities for internal improvements) and raising funds for the development effort. The second tier would consist of a professional executive (and staff) who would be responsible for organizing and implementing this policy. The third tier would consist of organizations, many of which currently exist, that would supply resources to the overall development effort. Since both public- and private-sector organizations have key roles to play in development activities, the development council should be set up as public-private partnerships whose sole mission is to promote economic development.

Members of the governing council from the public sector would include representatives of key governmental policymaking and administrative bodies with responsibilities for economic development–related matters, such as the city council, the Office of Recovery and Development Administration, the New Orleans Redevelopment Authority, and the Regional Planning Commission. Since the state also plays an important role in funding development programs, it should also have representation on the council. Members from the private sector should represent a diverse range of stakeholders, including the business, education, research, and facility-management sectors. Membership from the business community might include representatives from professional development agencies such as Greater New Orleans, Inc., the World Trade Center of New Orleans, and the Business Council, as well as individuals who represent the small-business community, such as members of the special-interest business chambers. Representatives from the other three sectors might include the president of one of the major universities, the director of a leading research center (perhaps in the medical research community), and the director of one of the major infrastructure facilities in the city (such as the head of the port or the airport). The objective in choosing the members for the development council is to gain both the perspective and the prestige that these representatives can bring to setting policy and raising funds for the effort.

The development council, in consultation with the executive, might also consider appointing task forces for each of the target industry clusters. These task forces would consist of a representative of the governing council, individuals from the various stakeholder communities, and the appropriate resource organizations in the third tier. A task force on tourism, for example, might include representatives from the convention and visitors bureau, the Tourism Marketing Corporation, the Louisiana Stadium and Exposition Board, the Audubon Commission, the Vieux Carre Commission, the Historic District Landmark Commission, and the city's arts and culture community. They would be tasked with recommending policies and priorities for action within each of the target industry areas.[19]

Implementing this organizational plan will require addressing two key issues. The first is the cost and funding of the effort. There are two sets of costs to consider: One relates to the operation of the economic development entity and its programs, and the other relates to the capital costs needed to rebuild and upgrade the city's infrastructure. Although most of the cities we examined did not report their annual operating costs, the range of those that did appears to be between $500,000 and $1 million. These funds are spent on salaries and various programs, including marketing costs. Perhaps a better gauge is the cost of running the city's Office of Recovery and Development Administration, which implements the Unified New Orleans Plan. The first year's budget for this office was $1.9 million and the proposed budget for fiscal year 2008 is $3.4 million.[20] Assuming that the economic redevelopment effort will require a similar operating budget, the operating costs for the agency administering this program should be between $2 million and $3 million per year.[21]

Like most other public-private partnerships, these costs should be borne by a combination of public and private sources. Currently, the city spends between $900,000 and $1.2 million annually on economic development–related programs. In addition, the city's economic development fund raised $5 million in FY 2007 and is projected to raise more than $8 million in 2008. A significant portion of the monies from this fund could be transferred to the budget of the new economic development agency. A similar sum should be raised from the private sector. In addition, funding might be sought from the private foundations that have already contributed to the recovery effort.

The second key issue relates to the operation of the new economic development entity and, in particular, how to ensure the cooperation of the various agencies and organizations that would constitute the third tier of the economic development effort. As noted earlier, these organizations have traditionally operated independently rather than cooperatively—each pursuing its individual agenda. Ensuring their cooperation in implementing a common plan and coordinating their efforts will be critical to the success of the economic development effort.

---

[19] One of our interviewees, Dan Storper, head of Putumayo World Music, made a number of recommendations for music development in the city that are likely to have broader applicability to the city's small-business sector. He noted that, despite the remarkable supply of musical talent in the city, New Orleans's music community has not connected to the music world well and has failed to take advantage of its strengths to build the city's music industry. Storper's suggestions included quarterly seminars for the music community that would bring in experts in sales, marketing, and promotion, as well as agents and promoters, to help the local industry better understand how to sell, market, and promote its product and to provide opportunities to network within the community. He also suggested hiring full-time promotion and marketing professionals to promote New Orleans and Louisiana music to national media and key markets (including foreign markets), creating a Louisiana culture office building to house local labels and arts businesses, setting up booths at major conferences for the industry to market the city's music and to promote the development of a music and culture trade conference in New Orleans, teaching business skills at the high school level, and expanding music and culture education at the elementary level.

[20] These figures are reported in the City of New Orleans proposed 2008 budget (see City of New Orleans, 2007).

[21] Without a detailed plan, it is impossible to estimate what the capital costs of the economic development plan would be. However, using the Bring New Orleans Back Commission estimate, the total capital costs (plus the costs of various state and federal tax incentives and whatever costs are paid directly by private firms) might be as much as $5 billion.

There are two ways in which this might be accomplished. The first is for the development agency to exercise control over the individual organizations' funding; the second is to exercise control over their expenditures. Since many of these organizations have independent sources of funding, it is unlikely that they would willingly cede their funding to the new agency or that the funding of the new agency would be sufficient to provide general support to each of these organizations from their own funds.[22] Changing the legislation that authorizes funding for the public agencies in this category would, no doubt, be politically difficult. On the other hand, it seems more feasible that these agencies might be required to obtain approval from the development agency before they spend that portion of their funds that is designated specifically for economic development activities. The executives of several of these organizations are likely to sit on the governing council, which will, in any case, have broad representation from key leaders from both the public and private sectors in the community. Thus, it might prove politically difficult for individual organizations to refuse to coordinate their activities with the economic development agency. However, implementation of this arrangement will require review of the various organizations' by-laws and municipal codes to determine what, if any, changes are required.

To ensure that narrow political interests do not subvert the operation of the new economic development agency (a frequent criticism of prior development programs), the executive responsible for implementing the development plan should be a professional with extensive experience in the development area. Moreover, since the development program will be a long-term effort, the executive should be appointed for a five-year term to ensure continuity of leadership. The executive should be appointed by the governing council and report to the council but should have considerable flexibility in implementing the development plan.

Implementation Component

The final component of an economic development program is the implementation phase. This phase typically addresses several issues, including its geographic focus (city versus region), the setting of priorities for the various activities (e.g., internal improvements), and creating the tools necessary to assess the program's progress and revise the plan as necessary. Each of these issues will be discussed in turn.

Geographic Focus

Most of the cities we examined adopted a regional rather than a city focus for their economic development programs in the belief that their primary competitors were other metropolitan regions rather than other communities within their own region. This is consistent with the fact that firms that are considering where to locate typically compare prospective locations at the metropolitan rather than community level (AngelouEconomics, 2007). However, we were told by several interviewees that there is often competition between New Orleans and the surrounding parishes and between New Orleans and the rest of the state. In fact, several respondents noted that firms are often drawn to the metropolitan area or to Louisiana by the attractions of New Orleans but then end up locating elsewhere in the state. The competition

[22] It is entirely likely, however, that the development agency would use some of its operating funds to provide funding for organizations to execute very specific support tasks.

among parishes and communities within the New Orleans metropolitan region may, in fact, have been intensified because the damage done by the storm and the level of recovery has varied so much from parish to parish.[23]

However, if industrial clusters rather than individual firms are the target, and even if the primary firm locates in a suburban area, the result is likely to benefit the city through increased employment opportunities for city residents and increased business for city firms that are part of the primary firm's supply chain. Moreover, several of the city's key industries, such as the port, oil and gas, and chemical industries, are distributed across the metropolitan area, and, thus, one area's gain may well benefit its neighbors. Indeed, cooperative agreements and planning across southeast Louisiana's ports and chemical industries should be explored.

Indeed, although interstate cooperation does not appear to be common practice in the Gulf Coast region, greater cooperation between New Orleans and surrounding cities and states in their economic planning and activities should be considered. This might include not only attempts to coordinate activities across such cities as New Orleans and Houston (both of which play key roles in the oil and gas industries), but also individual industries, such as between the Stennis Space Center in Mississippi and the NASA Michaud facility in New Orleans.

Setting Priorities

As noted earlier, a key issue that the development effort needs to address is how to set priorities among the many goals and activities embedded in the economic development plan. We also suggested several criteria to use in assigning priorities, including the city's need to have quick and visible successes, the potential synergies among alternative actions (e.g., a new cruise terminal would benefit both the port and the tourism industry), and a focus on those actions and policies that benefit the city as a whole rather than the interests of specific stakeholders.

Here, we consider an additional and even more important criterion: Namely, we suggest that the development council and the executive identify and place the highest priority on deal breakers, by which we mean those actions that are necessary conditions for the city's economic recovery. By their very nature, there will be few of these, but they are of paramount importance. The principal candidates for deal breakers are

- repair of the levees and a future level of flood protection that will prevent another Katrina from occurring
- improving the quality of life for the city's residents
- repairing the city's image.

Each of these actions is of paramount importance to the city's recovery. The prospect that the city will be left significantly underprotected against future storms acts as a powerful deterrent to investment in the city's economy for existing and new businesses and discourages displaced residents from returning and current residents from rebuilding their houses and neighborhoods. Similarly, improving the quality of life is necessary to replenish the city's current stock of residents and businesses with new residents and businesses—a necessary condition for

---

[23] The Orleans and St. Bernard parishes were hit particularly hard by the storm, and, by June 2006, their population and employment levels were still well below the pre-Katrina totals. By contrast, Jefferson Parish lost many fewer residents and has almost regained its pre-Katrina population total, and St. Tammany Parish has gained both population and employment since Katrina (Liu and Plyer, 2007). Indeed, a recent report suggests that competition among parishes within the New Orleans region has, in fact, intensified (AngelouEconomics, 2007).

full economic recovery. Although there are, as noted earlier, multiple aspects to this task, it will be particularly important for the city to solve the violent-crime problem that has made many New Orleanians fear for their safety and has contributed to the negative national perceptions of the city. Finally, the combination of pre-Katrina governance problems and the difficulties that New Orleans has faced during the recovery period have tarnished the city's reputation and severely limited its ability to attract new investment, business, and residents. Repairing that image will be critical to recovery. Although some might question the importance we place on the city's image to outsiders, one of the clearest lessons we draw from other cities' economic development programs is the emphasis they place on shaping perceptions of the city in the business community and among the general public. Indeed, in most of the cities we examined, their image to outsiders was viewed as a central element in their development strategy.

Although there are other potential candidates for deal-breaker status (e.g., the city's labor-pool problems, the way in which the city deals with business, the need to take quick action and not be viewed as simply undertaking another study and creating another set of plans),[24] we believe that the three issues we raise are the prime candidates. However, the critical need is for the city to identify what the potential deal breakers are and then setting out to address them first.

Setting Goals, Schedules, and Metrics for Assessing Progress

The final lesson to be drawn from other cities' economic development programs is the importance they attach to setting goals, scheduling to meet those goals, and identifying metrics for evaluating progress. As we discussed in Chapter Two, the priority assigned to different components of the plan should be translated into specific short-, medium-, and long-term goals, and those goals should then be mapped into a schedule (or multiple schedules, if task forces are used). The most important element here is the identification of metrics to evaluate progress toward meeting the goals. Finally, based on the assessment of progress using these metrics, the program needs to be modified accordingly.

The actual metrics used will vary with the phase of the development activity. In the design phase, for example, the metrics may consist of specific target industry plans and strategies to be produced by a specific date (identified in the goals and schedule) or a list of specific internal improvements to be made by a certain scheduled date. In the organizational phase, the measures used may refer to the establishment of target industry task forces and schedules for their completion of specific plans or fundraising targets to be realized. In the implementation phase, the metrics tend to be more concrete, such as passage of specific legislation, establishment of a one-stop shop for business permitting and licensing, or completion of capital improvement projects. These measures are typically designed to measure progress toward interim goals but can also reflect desired end states, such as the city reaching a specific population or employment level. Although the metrics used will vary with the specific plan and activity and regional or city context, the critical element is to have measurable goals against which to assess the development effort's progress and to make modifications as needed.

---

[24] Many of the interviewees emphasized that they, and the city's residents in general, wanted action now and that what the city did not need was another set of studies and plans for recovery.

CHAPTER FOUR

Summary of Key Findings and Recommendations

This chapter summarizes our key findings and highlights the principal recommendations of this report. The discussion of findings and recommendations is grouped into four sections. The first relates to the overall architecture of the city's development program, and the following three relate to the different components of that architecture (the design, organization, and implementation components). In each section, we begin by summarizing our key findings to provide a rationale for the recommendations that follow.

Overall Architecture

Our review of the experience of a range of different metropolitan regions suggests that effective economic development programs consist of three components: The first involves designing a comprehensive plan for the development program, the second establishes an effective organizational structure for the effort, and the third involves implementing the plan. This finding leads to our first recommendation.

Recommendation One: New Orleans should build its economic development program in three phases: a design phase, an organizational phase, and an implementation phase.

To gain widespread awareness of their economic development programs and to vest the various stakeholders that need to be involved in the program, the cities studied typically made a concerted effort to solicit a wide range of opinions about the goals of the program and to maintain the transparency of their operations. These characteristics are likely to be especially important for New Orleans's economic development program, given critiques of its prior development efforts, the controversy surrounding the various neighborhood redevelopment plans, and the city's reputation for cronyism.[1] This leads to our second recommendation.

Recommendation Two: The process of developing and implementing the city's economic development program should be both transparent (open to public scrutiny) and inclusive (incorporating input from a diverse range of stakeholders).

Although the recommendations that follow are consistent with the experience of successful economic development efforts in other regions, they are not a guarantee of a successful effort for New Orleans. The city must also be aware of national economic trends, responsive to the opportunities that these trends may present, and willing and able to adjust its plans as circumstances warrant and as the community expresses its desires for the future.

[1] Testifying to the importance of this issue, newly elected Governor Bobby Jindal has called a special session of the Louisiana legislature to deal with political corruption.

Design Phase

The design phase of the economic development programs we examined typically included four basic elements: a vision statement laying out the desired characteristics of the future economy; an assessment of the city's economic infrastructure that highlights its strengths and weaknesses; a description of the strategic approach of the effort, including its principal goals and targets; and a list of specific improvements to that infrastructure that are viewed as central to the plan.

Although vision statements are often cast in very general terms, the process of drafting them is important as a mechanism to publicize the development effort, involve a wide range of community stakeholders, and develop their support for the economic development program. Our interviews suggest that there is a wide range of opinion among individuals in New Orleans about the appropriate goals for the city's future and some skepticism as to whose interests will determine the outcome. This leads to our third recommendation.

Recommendation Three: The vision statement for a New Orleans economic development effort should involve consultation with a wide range of stakeholders to reflect the diversity of views and to make them aware of and vested in the development effort.

Deliberations about the appropriate strategies and targets for the development effort typically begin with an assessment of the strengths and weaknesses of the city's economic infrastructure. This practice reflects the fact that these elements of an effective development approach must be based on the local economy's current assets. Our review of New Orleans's economic base and its current assets revealed that the city's industrial base has been concentrated around four industrial clusters: a maritime cluster, consisting of shipping (the port) and shipbuilding and repair; an oil and gas cluster that includes production services and the petrochemical industry; a tourism cluster related to the city's unique facilities, festivals, and attractions; and a biomedical cluster that consists of health service and biomedical research industries. These clusters reflect the city's unique comparative advantages, most of which (e.g., its location on the Mississippi River and its extensive port facilities, its proximity to the nation's major oil and gas reserves in the Gulf of Mexico, its unique tourist attractions), will remain major advantages for New Orleans despite Katrina. On the other hand, the city's biomedical cluster, while based on a particularly strong set of facilities and a dominant position in the Gulf Coast region pre-Katrina, suffered major damage to facilities and professional workforce in the aftermath of the storm and will need to be repaired if the sector is to return to pre-Katrina capacity. Restoring the position of the biomedical industry will be a real challenge.

In addition, New Orleans has notable strengths in food processing, aerospace (the NASA Michaud facility), arts and entertainment (music and film production), and higher education, as well as promising clusters in such high-tech areas as information, biotech, and alternative energy. Finally, although devastated by Katrina, the small-business sector was a critical element of the city's economic infrastructure that provided 40 percent of the city's total employment before the storm and significant opportunities for minority entrepreneurs.

Our review of the city's economic base suggests that three groups of industries should be the targets of the city's economic recovery and development program, which leads to our next recommendation.

Recommendation Four: The city's economic development efforts should be targeted at three groups of industrial clusters. The first group consists of three core industries that have made or are making significant recovery post-Katrina—namely, the maritime, oil and gas,

and tourism industries. The second group consists of three groups of industries or sectors that are vital to the city's economy but have been slower to recover from the damage caused by the storm—namely the biomedical and food-processing industries and the city's small-business sector. These will require case-by-case evaluation. The third group consists of three clusters with solid pre-Katrina bases in the city and promising growth potential—namely, the arts and entertainment, information technology, and energy technology clusters.

In addition to its strengths, New Orleans's economic infrastructure has several weaknesses that must be addressed to enhance its economic development and revitalize its economic future. Many of these problem areas predate Katrina but have been exacerbated by the damage done by the storm. Probably the single most important physical infrastructure problem is the uncertain state of the levees and the flood-control system protecting the city from future storms. Despite the immediate repairs that have been made, the uncertainty surrounding the city's protection from future storms discourages much-needed investment, both by the city's existing businesses and residents and by potential new businesses and residents. In addition, the city's physical infrastructure needs major upgrading and repair. These improvements are needed not just to such basic facilities as streets, transportation networks, and utilities, but especially to the port and the city's biomedical facilities.

Other major problems for the city's economic infrastructure include the quantity and quality of its labor. Currently, there are not enough workers to fill the labor market's need. Problems in the city's workforce situation predate Katrina but have been substantially exacerbated since and will require not just an expansion in the size of the labor pool, but an upgrading of its quality. A third major area that needs improvement is the city's relations with the business sector. Members of the city's business community whom we interviewed repeatedly complained about the difficulties they faced in dealing with city government due to the complexity of and delays in the permitting and licensing process and the lack of a service orientation among city employees. This situation will pose special problems for the small-business sector, which will need special assistance to get reestablished.

Two other major problems that the city must address are upgrading the quality of life of its residents, particularly its violent-crime problem, and refurbishing the city's image. Despite its many charms, post-Katrina New Orleans is not an easy place to live. To remedy this situation, the city will need to address a major crime problem, upgrade the quality of its schools and health facilities, provide a sufficient supply of safe and affordable housing, and restore basic services to its neighborhoods. Finally, the New Orleans brand (i.e., the city's national image), which suffered pre-Katrina from perceptions of corruption and cronyism, has been further tarnished by the slow pace of recovery and additional indictments of city officials. Restoring the city's image is essential to securing the federal aid that the city needs and encouraging investment by businesses.

These weaknesses will need to be remedied, and the principal responsibility for doing so will rest with the city. This leads to our next recommendation.

Recommendation Five: New Orleans needs to recognize the importance of beginning its economic development programming by addressing the weaknesses in its economic infrastructure. Priority should be given to repairing and upgrading the city's infrastructure, expanding the supply and improving the quality of the city's labor force, improving the quality of the interaction between the city and the business community, upgrading the quality of life for the city's residents, and refurbishing the city's national image.

Our review of other cities' experiences indicates that they typically chose among three different general strategies for their economic development programs. Each approach has a distinct target and uses different selling points and marketing techniques. The first approach aims to recruit new industries by emphasizing low costs, tax incentives and rebates, and a business-friendly environment. It then aggressively markets these advantages to firms located in other areas. The second approach focuses on retaining and supporting the growth of industries and firms already located in the area. Its major selling points are an improved physical infrastructure, an abundant supply of qualified labor, and government policies that are supportive of local industries. It then focuses on making the needed internal improvements to the city's economic infrastructure. The third approach targets new potential growth industries, with a particular focus on high-tech industries. Its major selling points are a strong education and research sector and a quality of life that will attract highly skilled, relatively footloose workers. It then focuses on building up its educational and research capabilities and a quality of life that will attract a highly skilled labor force on the assumption that, if the area can attract such labor, the target industries will follow. The priorities that cities attach to these alternatives are typically based on the assessment of the area's economic infrastructure. These observations lead to our next recommendation.

Recommendation Six: New Orleans should attach the highest priority to the "retain-and-grow" approach and the second-highest priority to attracting high-tech growth industries.

Organizational Phase

To determine an effective organizational structure for New Orleans's economic development effort, we reviewed evaluations of the city's prior economic development efforts, as well as the characteristics of the organizational ecology of its development programs. These reviews supplemented our examination of other cities' approaches and our interviews in New Orleans.

A major evaluation of New Orleans's prior development efforts (Bureau of Governmental Research, 2004) highlighted several problems. First, the city's prior efforts lacked an overall strategy and comprehensive development plan and relied instead on an ad hoc approach that created the perception that decisions were based on political rather than economic considerations. Second, there was no single individual or organization in charge of decisionmaking. Instead, decisionmaking was spread among multiple organizations, with the result that were no clear priorities among alternatives. Third, no cost-benefit evaluation was conducted on individual projects or investments, so there was no way to determine whether they had met their objectives. Fourth, no attempt was made to assess the cumulative effects of individual decisions and their contribution to overall economic development.

Our review of the current organizational ecology indicates that, although there is a diverse range of public and private organizations at the state, regional, and local levels involved in economic development activities, many of the same problems remain. Specifically, these different organizations appear to have different emphases and agendas, represent diverse interests, and perform a wide range of functions. Moreover, since the leadership of these organizations tends to change as new administrations take office and individuals move on, there can be a lack of continuity of leadership. Although, in sum, they represent a valuable combination of resources and expertise, their efforts are not often coordinated, reflect different agendas and priorities, lack continuity of personnel, and fail to follow a common strategy or plan.

In combination, these observations lead to our next three recommendations, relating to the structure, roles, and funding of the proposed economic development agency. Based on the experiences of other cities, we suggest that New Orleans's economic revitalization program should have the following functional features: a comprehensive plan that all groups involved follow, a clear division of responsibility for major functions (e.g., policy, implementation, resources), continuity of leadership commensurate with the term of the effort, and coordination of the activities involved in implementing the plan. Correspondingly, we make the following recommendation.

Recommendation Seven: New Orleans's development agency should be led by a public-private partnership that consists of the following three components: a governing council comprised of leaders of the relevant public- and private-sector stakeholder organizations, which should have direct responsibility for setting policy for the development effort and raising funds for that effort; a professional executive staff led by the director of economic development, who would be appointed by the development council for a five-year term, report to the development council, and be responsible for implementing the plan and ensuring coordination of the development program's activities; and a resource group drawn from the various organizations and individuals currently involved in development that would provide needed resources and expertise, carrying out specific tasks at the direction of the executive director.

As we suggested, the executive director's ability to ensure cooperation from the organizations in the third component and to coordinate the development agency's activities would be enhanced by requiring him or her to specifically approve expenditures of funds that are directly related to economic development activities.

Recommendation Eight: The funding for New Orleans's public-private partnership should be provided by the public and private sectors.

Assuming that the costs of operating the agency will approximate the costs of running the city's Office of Recovery Development and Administration, the operating costs of the New Orleans economic revitalization partnership will be between $2 million and $3 million annually. Currently, the city is spending between $900,000 and $1.2 million per year on economic development activities. In addition, the city's economic development fund raised $5 million in FY 2007 and is projected to raise more than $8 million in 2008. A significant share of the proceeds from this fund should be dedicated to the new agency with a like amount provided by the private sector.

Recommendation Nine: The development council, in conjunction with the development executive, should consider appointing task forces that would be responsible for developing plans and implementation strategies for each of the target industry clusters.

Implementation Phase

A key lesson from other cities' development experiences is the importance of considering key implementation issues before embarking on an effort. The criticism of the city's prior efforts for failing to evaluate its programs' effectiveness, either individually or collectively, underscores the importance of such implementation issues. Based on the review of other cities' experiences, there are three specific issues that need to be addressed in the implementation phase. The first relates to the geographic focus of the development effort; specifically, should it concentrate on economic development at the city or regional level? The second concerns the criteria to use in setting priorities among the multiple goals of the effort. The final issue concerns how the effort measures progress and revises its plans accordingly.

Most of the cities we examined adopted a regional approach for their development programs in the belief that their major competitors were other regions rather than other communities within their own region. However, some of our interviewees in New Orleans told us that there was often competition between Orleans and the surrounding parishes and that prospective employers were sometimes drawn to the state by the attractions of New Orleans but ended up locating elsewhere in the state. Indeed, this interparish competition may have increased since Katrina. However, our review of the evidence suggests that both the city and the region would both benefit in the form of increased employment opportunities and business for suppliers from growth and the entry of new firms, whether that occurs in the city or surrounding areas. This leads to our next recommendation.

Recommendation Ten: The focus of New Orleans's development efforts should be the metropolitan region rather than the city exclusively.

A successful development effort will take time and involve a wide range of different activities. The list of improvements to the city's economic infrastructure discussed in this report, for example, included changes along several different dimensions, each of which will require a range of different actions. Moreover, the resources available to make these changes are limited. Correspondingly, a central issue for the development effort will involve setting priorities among competing goals. Thus, the development effort will need to determine what criteria to use to set those priorities. Although a variety of different criteria might be used, we have suggested several in Chapter Four.[2] What is paramount, however, is that criteria be chosen, that they be made public, and that their application be transparent. Following this guideline will go a long way to gaining credibility for the effort and gaining public acceptance for its actions. This leads to our next recommendation.

Recommendation Eleven: The development effort should set clear criteria for determining priorities among competing goals and actions, make those criteria public, and apply them in a transparent way. In particular, the development effort should place the highest priority on those actions that address conditions that might be considered "deal breakers."

The final element of the implementation phase is to establish a means to assess the progress of the development effort. As we have noted, one of the criticism of the city's past development program was its failure to assess the success of its efforts, either individually or collectively. As the experience of other cities demonstrates, remedying this problem requires benchmarks against which to assess the progress and effects of a development program. This involves translating priorities into goals, incorporating those goals into schedules, and developing metrics to measure performance in meeting these goals. Based on this progress, modifications can be made, as needed, to the implementation plan. This leads to our final recommendation.

Recommendation Twelve: The development plan should include a process for assessing progress. This will require setting goals, establishing schedules, and developing metrics for determining the progress and success of its efforts.

---

[2] Among the criteria suggested were the city's need for quick and highly visible successes, the possibility of synergy across goals and industries, the availability of funding, and, most importantly, identification of the deal breakers.

APPENDIX A

Development Agencies Researched

1. **Albuquerque, New Mexico**
   Albuquerque Economic Development
   851 University Blvd. SE, Ste. 203
   Albuquerque, NM 87106
   505-246-6200
   http://www.abq.org

2. **Asheville, North Carolina**
   Asheville Area Chamber of Commerce
   36 Montford Ave.
   Asheville, NC 28801
   828-258-6101
   http://www.ashevillechamber.org

3. **Atlanta, Georgia**
   Atlanta Development Authority
   86 Pryor St.
   Atlanta, GA 30303
   404-880-4100
   http://www.atlantada.com

4. **Birmingham, Alabama**
   Birmingham Metropolitan Development Board
   500 Beacon Parkway West
   Birmingham, AL 35209
   205-942-7284
   http://www.mdb.org

   City of Birmingham, Office of Economic Development
   710 North 20th St.
   Birmingham, AL 35203
   205-254-2799
   http://www.informationbirmingham.com/economic.htm

Birmingham Business Resource Center
1500 1st Ave. North, Ste. B108
Birmingham, AL 35203
205-250-6380
http://www.bbrc.biz

5. **Charleston, South Carolina**
Charleston Regional Development Alliance
5300 International Blvd., Ste. 103A
North Charleston, SC 29418
843-767-9300
http://www.crda.org

City of Charleston, Department of Economic Development
75 Calhoun St., 3rd Floor
Charleston, SC 29401
843-724-3796
http://www.charlestoncity.info

6. **Houston, Texas**
Greater Houston Partnership
1200 Smith St., Ste. 700
Houston, TX 77002
713-844-3600
http://www.houston.org

7. **Jackson, Mississippi**
Greater Jackson Alliance
P.O. Box 3318
Jackson, MS 39207
601-948-3111
http://www.metrojacksoneda.com

MetroJackson Chamber of Commerce
P.O. Box 22548
Jackson, MS 39225
601-948-7575
http://www.metrochamber.com

8. **Jacksonville, Florida**
Jacksonville Economic Development Commission
1 West Adams St., Ste. 200
Jacksonville, FL 32202
904-630-1858
http://www.coj.net/Departments/Jacksonville+Economic+Development+Commission/default.htm

Cornerstone Regional Development Partnership
Jacksonville Regional Chamber of Commerce
3 Independent Dr.
Jacksonville, FL 32202
904-366-6680
http://www.expandinjax.com

9. **Miami, Florida**
Beacon Council
80 SW Eighth St., Suite 2400
Miami, FL 33130
305-579-1300
http://www.beaconcouncil.com

Greater Miami Chamber of Commerce
1601 Biscayne Blvd.
Miami, FL 33132
305 350 7700
http://www.greatermiami.com

City of Miami, Department of Economic Development
444 SW 2nd Ave., 3rd Floor
Miami, FL 33130
305-416-1435
http://www.ci.miami.fl.us/economicdevelopment

10. **Nashville, Tennessee**
Nashville Area Chamber of Commerce
211 Commerce St., Ste. 100
Nashville, TN 37201
615-743-3024
http://www.nashvilleareainfo.com

11. **Virginia Beach–Norfolk–Newport News–Hampton, Virginia**
Hampton Roads Economic Development Alliance
500 Main St., Ste. 1300
Norfolk, VA, 23510
757-627-2315
http://www.hreda.com

12. **Orlando, Florida**
Metro Orlando Economic Development Commission
301 East Pine St., Ste. 900
Orlando, FL 32801
407-422-7159
http://www.orlandoedc.com

City of Orlando, Business Development
One City Commons
400 South Orange Ave.
Orlando, FL 32801
407-246-2121
http://www.cityoforlando.net/executive/eco_dev/vision.htm

13. **Richmond, Virginia**
Greater Richmond Partnership, Inc.
901 E. Byrd St.
Richmond, VA 23219
804-643-3227
http://www.grpva.com

14. **San Antonio, Texas**
City of San Antonio Economic Development
Frost Bank Tower
100 West Houston St., Ste. 1900
San Antonio, TX 78205
210-207-8080
http://www.sanantonio.gov/edd/index.asp

San Antonio Economic Development Foundation
P.O. Box 1628
602 E. Commerce St.
San Antonio, TX 78296
210-226-1394
http://www.sanantonioedf.com

15. **San Diego, California**
San Diego Regional Economic Development Corporation
530 B St., 7th Floor
San Diego, CA 92101
619-234-8484
http://www.sandiegobusiness.org

16. **Savannah, Georgia**
Savannah Economic Development Authority
31 Hutchinson Island Rd., 4th Floor
Savannah, GA 31421
912-447-8450
http://www.seda.org

17. **Winston-Salem, North Carolina**
    Piedmont Triad Partnership
    7025 Albert Pick Rd., Ste. 303
    Greensboro, NC 27409
    336-668-4556
    http://www.piedmonttriadnc.com

APPENDIX B

Alphabetical List of Interviewees

Donna Addkison, principal, Sun Consulting, LLC

Arnold B. Baker, founder and CEO, Baker Ready Mix

Camille Barnett, senior managing consultant, Public Financial Management

Edward Blakely, executive director, City of New Orleans Office of Recovery and Development Administration

Donald "Boysie" Bollinger, chair and CEO, Bollinger Shipyards, Inc.

Hal Brown, partner, Paladin Capital Group, and chair, Horizon Initiative

James Carter, member, New Orleans City Council

Cynthia Ceballos-Hymel, Ceballos Legal Consulting, LLC, and president, Hispanic Chamber of Commerce of Louisiana

John Edwards, associate professor of economics, Tulane University

Lolis Eric Elie, columnist, *Times-Picayune*

Laurance "Larry" Eustis III, chair, Eustis Mortgage

Ron Forman, CEO Audubon Nature Institute

N. Sundiata Haley, Haley and McKee, LLC

Stacy Head, member, New Orleans City Council

Jill Hickson, chief of staff, Office of Arnie Fielkow, New Orleans City Council

William H. Hines, managing partner, Jones, Walker, Waechter, Poitevent, Carrère and Denègre, LLP

Janet Howard, president and CEO, Bureau of Governmental Research

Robin Keegan, economic and workforce development policy director, Louisiana Recovery Authority

Jay LaPeyre, president, Laitram, LLC, and chair, Business Council of New Orleans

Don Marshall, executive director, New Orleans Jazz and Heritage Festival and Foundation, Inc.

Robert W. Merrick, chair and CEO, Latter and Blum, Inc.

Shelley Midura, member, New Orleans City Council

Ivan Miestchovich, Jr., associate profesor of finance and director, Center for Economic Development, University of New Orleans

R. King Milling, president, Whitney Bank

Steve Molnar, director of regional business development, Greater New Orleans, Inc.

Patricia Morris, research analyst, Bureau of Governmental Research

Tuan Nguyen, interim business development director, MQVN Community Development Corporation

Michael J. Olivier, secretary, Louisiana Department of Economic Development

Tom O'Malley, vice president and director, Gulf Coast Revitalization Program, AFL-CIO Investment Trust Corporation

Kevin Pollard, managing partner and president, GlobalSolve Management Services

Arthur Pulitzer, cofounder, Horizon Initiative

Sean Reilly, COO and president, Lamar Advertising Company, Outdoor Division, and board member, Louisiana Recovery Authority

Patricia Robinson, deputy director of development, City of New Orleans Office of Recovery and Development Administration

William C. Rouselle, CEO, Bright Moments

Eugene J. Schreiber, managing director, World Trade Center of New Orleans

Michael J. Siegel, executive vice president and director of leasing, Corporate Realty, Inc.

Dan Storper, founder and CEO, Putumayo World Music

Kurt M. Weigle, president and CEO, Downtown Development District

George Wentz, partner, Baldwin, Haspel, Burke and Murphy, and cofounder, Horizon Initiative

Tim Williamson, cofounder and president, Idea Village

Bibliography

AngelouEconomics, *Forward Charleston: Targeted Economic Development and Marketing Strategy, Report 2: Economic Development Strategy*, Charleston, S.C., 2005.

———, *Greater New Orleans Regional Market Assessment*, New Orleans, La.: Greater New Orleans Inc., August 2007.

Birch, Eugenie, and Susan Wachter, *Rebuilding Urban Places After Disaster: Lessons from Hurricane Katrina*, Philadelphia, Pa.: University of Pennsylvania Press, 2007.

BNOB—*see* Bring New Orleans Back Commission.

Bring New Orleans Back Commission, *Post-Katrina Economic Redevelopment Plan*, New Orleans, La., updated May 25, 2006. As of November 20, 2007:
http://www.bringneworleansback.org/Portals/BringNewOrleansBack/Resources/Mayors%20Rebuilding%20Plan%20Final.pdf

Bureau of Governmental Research, *On the Right Track? New Orleans Economic Development in Perspective*, New Orleans, November 2004. As of November 20, 2007:
http://www.bgr.org/BGR%20in%20the%20News/BGR%20Reports/Economic%20Development/Economic%20Development%20Report%2011-04.pdf

Charleston Regional Development Alliance, homepage, undated. As of November 20, 2007:
http://www.crda.org

City of New Orleans, Division of Housing and Neighborhood Development, "Neighborhood 1," Web page, undated. As of November 20, 2007:
http://www.cityofno.com/Portals/Portal49/portal.aspx?portal=49&tabid=1

———, *City of New Orleans 2008 Budget*, New Orleans, La., 2007. As of November 20, 2007:
http://www.cityofno.com/Resources/2008mayorsbudget.pdf

City of San Antonio, Economic Development Department, "San Antonio, Inclusive," homepage, undated. As of November 20, 2007:
http://www.sanantonio.gov/edd/collaboration/sainc/sainc_index.asp

DADCO Consulting Services, Inc., *The Power of Partnerships: Clusters and Collaborative Economic Development in Southeast Louisiana*, Shreveport, La., 2002.

Florida, Richard, *The Rise of the Creative Class*, New York: Perseus Books, 2002.

Fruth, William H., *Economic Strength Rankings—2007*, Palm Beach, Fla.: Policom Corporation, 2007. As of November 20, 2007:
http://www.policom.com/EconomicStrength2007.pdf

Greater Louisville Inc., High Impact Program, homepage, 2007. As of November 20, 2007:
http://www.greaterlouisville.com/highimpactprogram

Greater New Orleans, Inc., homepage, 2007. As of November 20, 2007:
http://www.gnoinc.org

Liu, Amy, and Allison Plyer, *New Orleans Index: A Review of Key Indicators of Recovery Two Years After Katrina*, Washington, D.C., and New Orleans: Brookings Institution and Greater New Orleans Community Data Center, August 2007. As of November 20, 2007:
http://www.brookings.edu/reports/2007/~/media/Files/Metro%20Simple%20Pages/ESNOLAIndexAug07.pdf

Louisiana Economic Development Council, *Louisiana: Vision 2020, 2003 Update: A Master Plan for Economic Development*, Baton Rouge, La., 2003. As of November 20, 2007:
http://vision2020.louisiana.gov

Louisiana Economic Development Department, homepage, 2007. As of November 20, 2007:
http://www.lded.state.la.us

Mann, Charles C., "The Long Strange Resurrection of New Orleans," *Fortune*, August 21, 2006, pp. 86–109. As of November 20, 2007:
http://money.cnn.com/magazines/fortune/fortune_archive/2006/08/21/8383661/index.htm

McCarthy, Kevin, D. J. Peterson, Narayan Sastry, and Michael Pollard, *The Repopulation of New Orleans After Hurricane Katrina*, Santa Monica, Calif.: RAND Corporation, TR-369-RC, 2006. As of November 20, 2007:
http://www.rand.org/pubs/technical_reports/TR369

MDC, Inc., *The State of the South, 2002: Shadows in the Sunbelt, Revisited*, Chapel Hill, N.C., September 2002. As of November 20, 2007:
http://www.mdcinc.org/docs/sos_02.pdf

Memphis and Shelby County Office of Economic Development, Web page, undated. As of November 20, 2007:
http://www.dobizinmemphis.com

MetroVision School-to-Career Partnership, *The Power of Partnerships: Clusters and Collaborative Economic Development in Southeast Louisiana*, New Orleans, La., 2001.

Mt. Auburn Associates, *Louisiana: Where Culture Means Business*, Baton Rouge, La.: State of Louisiana Office of the Lieutenant Governor, Department of Culture, Recreation and Tourism, July 31, 2005. As of November 20, 2007:
http://www.crt.state.la.us/CulturalEconomy/MtAuburn/culturaleconomyreport.htm

New Orleans Chamber of Commerce, homepage, undated. As of November 20, 2007:
http://www.neworleanschamber.org

New Orleans Redevelopment Authority, homepage, undated. As of November 20, 2007:
http://www.noraworks.org

Pike, Jennifer, *Spending Federal Disaster Aid: Comparing the Process and Priorities in Louisiana and Mississippi in the Wake of Hurricanes Katrina and Rita*, Baton Rouge, La.: Nelson A. Rockefeller Institute of Government and Public Affairs Research Council of Louisiana, 2007. As of November 20, 2007:
http://www.rockinst.org/publications/subpage.aspx?id=12270

Regional Loan Corporation, homepage, 2006. As of November 20, 2007:
http://www.rlcsbidco.com

Regional Planning Commission, New Orleans, La., homepage, undated. As of November 20, 2007:
http://www.norpc.org

Regional Technology Strategies, Inc., *Building a Statewide System to Foster University Technology Development and Technology Commercialization, Phase One: A Strategic Plan for the Research Commercialization and Educational Enhancement Program*, Carrboro, N.C., February 2007. As of November 20, 2007:
http://laregents.org/www2/main/RTS%20Phase1%20Strategic%20Plan.pdf

Terrell, Dek, and Ryan Bilbo, *A Report on the Impact of Hurricanes Katrina and Rita on Louisiana Businesses: 2005Q2–2006Q2*, Baton Rouge, La.: E. J. Ourso College of Business, Louisiana State University, 2006. As of November 20, 2007:
http://lra.louisiana.gov/assets/LouisianaBusinessRecoveryReport030807final.pdf

Thomas, G. Scott, "Where Paychecks Are Growing Fastest," *Bizjournals*, June 4, 2007. As of November 20, 2007:
http://philadelphia.bizjournals.com/edit_special/52.html

Tortorano, David, and Gloria Tortorano, *Gulf Coast I-10 Research Directory 2006*, Morrisville, N.C.: Lulu Enterprises, 2006.

Unified New Orleans Plan, *Citywide Strategic Recovery and Rebuilding Plan*, New Orleans, April 2007. As of November 20, 2007:
http://unifiedneworleansplan.com/home3/section/136/city-wide-plan

University of New Orleans, Division of Business and Economic Research, *Metropolitan Report: Economic Indicators for the New Orleans Area*, Vol. 18, No. 1, August 2007. As of November 20, 2007:
http://www.lded.state.la.us/uploads/pdf/Metro%20Report%20Katrina%202nd%20Anniversay%20Aug07.pdf

UNOP—*see* Unified New Orleans Plan.

Urban Land Institute, *New Orleans Louisiana: A Strategy for Rebuilding*, Advisory Services Program Report, Washington, D.C., November 12–18, 2005. As of November 20, 2007:
http://www.uli.org/AM/Template.cfm?Section=Home&CONTENTID=80035&TEMPLATE=/CM/ContentDisplay.cfm

U.S. Census Bureau, "2006 American Community Survey Data Profile Highlights," undated(a). As of September 24, 2007:
http://factfinder.census.gov/servlet/ACSSAFFFacts

———, American FactFinder, DP-1, "General Population and Housing Characteristics: 1990," undated(b). As of September 24, 2007:
http://factfinder.census.gov/servlet/QTTable?_bm=y&-geo_id=04000US72&-qr_name=DEC_1990_STF1_DP1&-ds_name=DEC_1990_STF1_&-redoLog=false&-CONTEXT=qt

———, American FactFinder, DP-1, "Profile of General Characteristics: 2000," undated(c). As of September 24, 2007:
http://factfinder.census.gov/servlet/QTTable?_bm=y&-geo_id=01000US&-qr_name=DEC_2000_SF1_U_DP1&-ds_name=DEC_2000_SF1_U&-redoLog=false&-CONTEXT=qt

———, American FactFinder, DP-3, "Profile of Selected Economic Characteristics: 2000," undated(d). As of September 24, 2007:
http://factfinder.census.gov/servlet/QTTable?_bm=y&-geo_id=01000US&-qr_name=DEC_2000_SF3_U_DP3&-ds_name=DEC_2000_SF3_U&-_sse=on

U.S. Department of Labor, Bureau of Labor Statistics, "States and Metropolitan Areas at a Glance," undated. As of September 6, 2007:
http://www.bls.gov/eag